CMOS High Efficiency On-chip Power Management

T0137773

ANALOG CIRCUITS AND SIGNAL PROCESSING SERIES

Consulting Editor: Mohammed Ismail. *Ohio State University*

For further volumes:
http://www.springer.com/series/7381

John Hu • Mohammed Ismail

CMOS High Efficiency On-chip Power Management

Springer

John Hu
Analog VLSI Lab
Department of Electrical and Computer
Engineering
The Ohio State University
Columbus, OH
USA
Hu.193@osu.edu

Mohammed Ismail
Analog VLSI Lab
Department of Electrical and Computer
Engineering
The Ohio State University
Columbus, OH
USA
ismail@ece.osu.edu

On leave as ATIC Professor
Chair and Campus Director
The Khalifa University of
Science, Technology and
Research (KUSTAR), UAE

ISBN 978-1-4614-2939-5   ISBN 978-1-4419-9526-1 (eBook)
DOI 10.1007/978-1-4419-9526-1
Springer New York Dordrecht Heidelberg London

Printed on acid-free paper

Springer is part of Springer Science+Business Media (www.springer.com)

To my parents, Hu Heping and Xiao Lijun.

To the memory of my late father, Ismail A. Elnaggar.

Preface

The world we live in today faces two fundamental challenges: the need to search for renewable energy sources as replacements for depleting fossil fuels, and the urgent necessity to reduce green house gas emissions that are currently threatening the sustainability of the common and only habitat of human beings – the earth – through global warming.

With wind and solar energy harvesting still facing major hurdles in consistency and reliability, and the safety of nuclear energy production called into serious questions after the tragic Fukushima Daiichi nuclear disaster, improving energy efficiencies in all aspects is identified as probably the most effective way to alleviate both crises in the short term.

Parallel to the energy and environmental crises at the megawatt level, the consumer electronics industry is experiencing a similar but miliwatt level energy crisis of its own. Consumers appreciate the multimedia and connectivity a portable device can provide, but continue to demand more functionality, versatility, and most importantly longer battery runtime in the same compact size.

Although there has been great progress in developing low cost, high energy density battery technologies, enhancing the energy efficiency of the electronic circuits and systems through aggressive and innovative power management is still considered as the most effective and extremely necessary method for the electronics vendors to meet consumers' needs for a longer battery life and develop products that will be competitive in the market. Better efficiencies will also reduce heat, which can be very undesirable for end users.

It is because of these reasons that there is a renaissance of interest in power electronics within the industrial and research communities. Technologies of power electronics date back to days when power grids were first laid and electric plants were created. It is only recently that they see brand new applications in a wider field of interest: smart grid, power management for consumer electronics, and biomedical applications, just to name a few.

This book deals with the subject matter of power management IC design, or integrated power electronics, as a response to the growing need for energy-efficient electronics. It is an emerging field that has lately grown into an area of its own in parallel with conventional analog, digital, RF, and mixed-signal IC designs. This field is unique in that it requires considerable amount of understanding of power converters and principles. It is also very challenging as it leverages VLSI techniques for implementation.

In particular, this is the first monograph that addresses power management IC design with an emphasis on high efficiency and full on-chip integration. High efficiency is essential to extend battery life and reduce heat. Maximum on-chip integration with fewer external components is the growing trend, as the physical size, bill-of-materials, and manufacturing cost all point to a system-on-chip solution in a mainstream CMOS process.

This book is divided into two parts with four chapters. Part 1 (Chaps. 1 and 2) presents the system point of view on power management in a green electronic system. Part 2 (Chaps. 3 and 4) goes through details of circuit level power management IC design.

Chapter 1 introduces the concept of green electronics in face of the power crisis in portable consumer electronics. It envisions the structure and necessary components of such a system and identifies power management blocks as the bottle neck for overall efficiency improvement. As an introduction, it also describes the uniqueness of power management IC design and the rationale behind the study thereof.

Chapter 2 discusses power management at the system level. A holistic approach that involves system level software, SoC architecture, and silicon IPs is presented. Meanwhile, the importance of sleep mode operation for portable and battery-powered applications is discussed. The runtime extension using sleep-mode efficiency IPs is quantized, and the circuit level design of these sleep-mode efficient IPs becomes a theme for the second part of the book.

Chapter 3 starts dealing with power management IC design by analyzing linear regulators, especially the low drop-out (LDO) topology. Key performance parameters of LDOs are listed. Their design challenges are also explained, and existing performance enhancement methods are reviewed. In the quest for full on-chip integration, designs of external capacitor-free LDOs are presented. A novel sleep-mode ready, area-efficient capacitor- free LDO is proposed with testing methods and measurement results presented.

Chapter 4 moves on to highly efficient power management IC design, represented by switching converters. The light-load efficiency of these converters is found to be insufficient, and the root cause for the efficiency roll-off is identified. Existing light-load efficiency boosting techniques are then discussed, followed by a proposal of a long-sleep model (LSM). A design example of a light-load efficient DC–DC buck converter using the LSM is presented. The characteristics, implementation, implication, and novelty of the LSM are studied thoroughly.

This book will serve as a reference for analog and power management IC design engineers in industry, as well as graduate students conducting research in high efficiency electronics, power management, and analog IC design in CMOS technologies. It will also be useful for test engineers, project leaders, design managers, and individuals in marketing and business development.

This book has its root in the Ph.D. dissertation of the first author at the Analog VLSI Laboratory at the Ohio State University. We would like to thank all those who supported us, including our colleagues in the Analog VLSI Laboratory and the Electro science Laboratory at the Ohio State University, at First-pass Technologies in Dublin, Ohio, and from Battery Management Solutions and Low Power RF divisions of Texas Instruments. Last but not least, we would like to thank our families for their continued support and understanding.

Dallas, Texas   John Hu
Columbus, Ohio   Mohammed Ismail

Contents

Part I
Green Electronics and Power Management

Chapter 1
Green Electronics

The last few decades have seen climate disruption unprecedented over the recent millenia. The trend toward global warming is virtually certain: the earth surface has increased by about 0.74°C since 1906 (Vit 2010), and there is a strong consensus among scientists that this is a result of the emission of greenhouse gases, preeminent of which is CO_2, mainly originating from fossil fuel burning. Aside from the negative environmental impact, fossil fuels are non-renewable resources that take millions of years to form, but are now depleting at a much faster speed than new ones are being made. As a result, the world is facing an unprecedented challenge to switch from fossil fuel, carbon-emission driven growth to an eco-friendly, low carbon economy to meet the energy need as well as mitigating the environmental consequences (NRE 2010).

Next to the replacement of fossil fuels by alternatives, such as solar, wind, geothermal, biomass, and so forth, there is also increasing awareness that scrutiny of existing technology and applications can have an immediate, and profound impact (de Vries 2010). Energy efficiency (EE), which is to reduce the amount of energy required to provide products and services, is often considered as the parallel pillar to renewable energy (RE) in a sustainable energy economy (EER 2010). Parallel to the large-scale energy grand challenge, portable and battery-powered products are facing a similar power crisis: consumers who appreciate the multimedia experience and ubiquitous connectivity in an increasingly compact size, are unwilling to sacrifice, and even demand longer battery runtime for new generations of devices.

This book, therefore, is focused on improving the energy efficiency of electronic products, especially portable and battery-powered ones, using advanced power management and very-large-scale integration (VLSI) design techniques, which will not only effectively reduce the power consumption and extend the battery life for current consumer products, but also enable and assist future clean energy generation and distribution, leading to a new era of "Green Electronics" in the future renewable-energy economy.

J. Hu and M. Ismail, *CMOS High Efficiency On-chip Power Management*, Analog
Circuits and Signal Processing, DOI 10.1007/978-1-4419-9526-1_1,
© Springer Science+Business Media, LLC 2011

1.1   The Concept of Green Electronics

The term green electronics has been used to refer to a number of initiatives. From the perspective of manufacturing, "green" has been used to refer to a design and manufacturing process that is environmentally friendly, which does not create or involve the use of hazardous materials and chemicals (Shina 2008). The motivation is that consumers today are increasingly more environmentally aware and that they are willing to pay a little extra for the same quality and performance if the products they purchase are green (Saphores and Nixon 2007).

From the perspective of recycling and waste management, "green" may also refer to advanced product designs that have less impact on the environment by using cleaner materials, having a longer product life, and reducing the electronic waste (e-waste) generated (Ogunseitan et al. 2009). It may also refer to public programs that encourages reuse and recycle of electronics (e-Cycle), such as computers and cell phones (EPA 2010). The rationale is that electronic devices consist of valuable resources such as precious metals, copper, and engineered plastics, all of which require considerable energy to process and manufacture. E-Cycle recovers valuable materials, conserves virgin resources, and results in lower environmental emissions (including greenhouse gases) than making products from virgin materials.

From the perspective of sustainable practices, "green electronics" is often used interchangeable with "green computing", which refers to the environmentally responsible use of computers and related resources, such as powering down the CPU and all peripherals during extend period of inactivity, powering up energy-intensive peripherals such as laser printer according to need, and using power management features for hard disks and displays (Murugesan 2008). For consumers planning to purchase new electronic devices, there are various tools to help them make the right choices to save money and protect the environment. Energy Star, for instance, is a joint program of U.S. Environmental Protection Agency (EPA) and the U.S. Department of Energy (DOE) started in 1992 and designed to identify and promote energy-efficient products to reduce greenhouse gas emissions (Ene 2010).

Finally, from the perspective of electronic design, "green" refers to the innovative designs and techniques that improve energy efficiency. In addition to the benefit of reducing energy consumption, carbon emission, and e-waste disposal, energy efficiency reduces the cost and increases the appeal of the products. In high performance computing applications such as data centers, power and cooling cost has increased by 400% over the past decade, and these costs are expected to continue to rise (as seen in Fig. 1.1). In some cases, power cost accounts for 40–50% of the total data-center operation budget (Filani et al. 2008). In portable and battery-powered applications like laptops, cellphones, and PDAs, higher energy efficiency translates to longer runtime for the same rechargeable battery, which improves

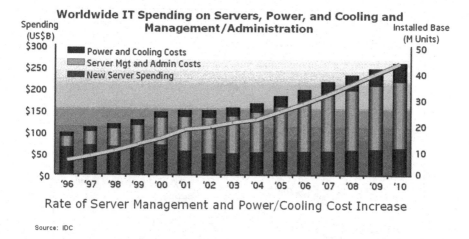

Fig. 1.1 Rate of server management and power/cooling cost increase (courtesy of Filani et al. (2008))

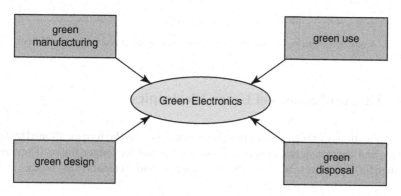

Fig. 1.2 A holistic approach toward green electronics

consumer quality-of-experience (QoE) by requiring fewer recharges (Reiter 2009). It also reduces heat dissipation, which eliminates the need for large heat sinks, enabling smaller, thinner, and more compact electronic products.

This book mostly uses the term "green electronics" in its last interpretation. In other word, it focuses on the technical challenges and solutions in designing energy efficient electronic products. However, it is important to realize that the goal of green electronics can only be achieved with a joint effort from all perspectives, and that the combination of green design, green manufacturing, green computing, and green disposal forms a holistic approach to green electronics in a sustainable economy (Murugesan 2008) (as seen in Fig. 1.2).

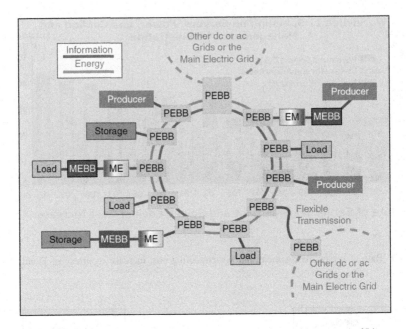

Fig. 1.3 A possible scenario of future power system based on smart grid (courtesy of Liserre et al. (2010))

1.2 The Applications of Green Electronics

The applications of green electronics go beyond relieving the fore mentioned power, energy, and environmental crises in the near term and the longer future. They may also benefit various research in industrial, science, and medical (ISM) fields.

1.2.1 Smart Grid

A major potential application of green electronics is in the future smart grid. The 2007 Energy Independence and Security Act defined the term "smart grid" as the modernization of the electricity delivery system that monitors, protects, and automatically optimizes the operation of its interconnected elements – from a central and distributed generator through a high-voltage transmission network and distribution system, to industrial users and building automation systems, and to end-use consumers and their thermostats, appliances, other handheld devices, and even electric vehicles (Schneiderman 2010).

A possible scenario of future power system based on smart grid technologies can be modeled by two concentric circles, as illustrated in Fig. 1.3. The outer circle represents energy flow and the inner circle models information flow over communication networks (Liserre et al. 2010). Power electronics building blocks (PEBB) and mechanical building blocks (MEBB) are needed in a number of energy "hubs", which manage multiple energy carriers (e.g., electricity, natural gas, and district heating) and transform part of the energy flow into another form of energy.

As a result, a variety of green electronic circuits would be needed. To ensure efficient energy flow and flexible interconnection of the different smart grid players (producers, energy storage systems, and loads), highly efficient power converters in the form of high voltage DC and flexible AC converters, bidirectional energy conversion structures adopting pulse-width modulation (PWM) technology and other control algorithms are essential (Liserre et al. 2010). To better match energy demand, avoid excessive load peaks, and coordinate between producers and consumers, communication and information technologies would play a critical role (Ipakchi and Albuyeh 2009). Information exchange in the upper grid levels is usually covered by existing communication networks such as on fiber-optic links that are installed in parallel to the high voltage grid, but medium and low voltage interconnection remains to be developed. Public telecommunication network (GSM, GPRS, UMTS), wireless network (WLAN, WiMAX), or powerline communication systems are all potential candidates, which all need large numbers of high efficiency communication electronics. In addition, power generation at the producers also opens up a significant demand for digital signal processing (DSP) and microcontroller (MCU) technologies. For example, in wind power generation, optimizing the angle of the blades of wind mill would require some sophisticated and signal processing and computing (Schneiderman 2010). This is a good example of green electronics minimizing energy usage in various other applications and assisting clean energy generation.

1.2.2   Wild Life Monitoring

Another possible application of green electronics is in wild life monitoring and research. In ecology and social biology, there are huge interests in observing long-distance migration, inter-species interaction, and nocturnal behavior. For example, Central Kenya is becoming more densely populated with smaller landholdings and more crop acreage. How fences and human presence affect large scale (tens of kilometers) zebra migration is still unknown (Zeb 2004).

Existing technologies applied toward wild life tracking are surprisingly primitive or overly expensive. They include attaching VHF transmitters to the animals and use flyovers to detect locations and using commercial GPS and satellite uploads to track locations. (Martonosi 2002) Recently, wireless ad-hoc networks that automatically discover neighbors, self-organize, and perform peer-to-peer data routing was also proposed (Juang et al. 2002). However, a common limitation of these methods is power management, which includes the constraint of the battery power given

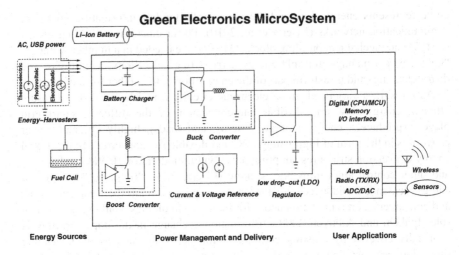

Fig. 1.4 Block diagram of a green electronics micro system

the weight limit, and the energy efficiency of the hardware platforms and network algorithms. Future green electronics with improved energy efficiency would prolong the tracking period between each battery recharge, increase the reliable signal transmission distance by allowing more complex radio communication, and enable long-term non-invasive observations.

1.3 Green Electronics Systems

With various existing and emerging applications of green electronics, it is impossible and unnecessary to define a universal structure or system architecture. However, it remains meaningful to outline a possible solution targeting a selective subset of applications as a means to identify the various challenges. Figure 1.4 shows such a system: a portable, battery-powered or self-sustainable, low-power green electronic microsystem.

The green electronics micro system can be divided into three parts: energy sources, user applications, and power management. Energy sources may include primary (non rechargeable) or secondary (rechargeable) batteries, 110 V AC wall power, 5 V DC USB power, and non-conventional energy sources such as energy harvesters and fuel cells. Many portable and handheld electronics today power off a single-cell or multi-cell Li-Ion batteries, and they would recharge the battery pack whenever an AC adapter or USB cable is connected (TPS 2010). The main problem with this configuration is that the total amount of energy available for the system is limited, and that this total amount is reduced overtime as the battery ages. Energy harvesters, on the other hand, are able to scavenge ambient kinetic (Torres and Rincon-Mora 2010), thermal (Carlson et al. 2010), solar (Guilar et al. 2009), and RF (Le et al. 2008) energy and convert them into electric power. A boost converter (Carlson et al. 2010; Dayal et al. 2010) that up converts the low DC or rectifies the AC output voltages (Yoo et al. 2010; Ramadass and Chandrakasan 2010) of these harvesters are needed in the system. A third possible energy source is a fuel cell (Meehan et al. 2010; Kim and Rincon-Mora 2009), which generates electric current through reactions between a fuel and an oxidant. But due to the low output voltage of fuel cells, boost converters, charge pumps, or a combination of both will be needed (Meehan et al. 2010).

The second part of the green electronics systems is user applications. In order to provide maximum functionality and add value to different applications, the system envisioned in Fig. 1.4 includes the following key components:

- *Intelligence.* Electronic intelligence is often realized through embedded computing in the form of microcontroller (MCU) or central processing unit (CPU). Random access memory (RAM) and various forms of read-only memory (ROM) are also needed to store the corresponding algorithms and processed data.
- *Communication.* Green electronics in future smart grid or wild life monitoring applications would also require wired or most likely wireless communication capabilities to send useful information and coordinate operation. The communication protocol may not be advanced and the data rate need not be high, but this feature is essential for mobility and networking. As a result, some form of radio transceiver (TX/RX) is assumed present.
- *Sensing.* Another important component of green electronics is its sensing and control circuits. In future homes connected to the smart grid, temperature, humidity, and motion detectors would be needed to automatically control air conditioning and other power utilities. Analog-to-digital (A/D) and digital-to-analog (D/A) converters will be required to sample, measure, and adjust various real world analog signals.

The third part of the system is power management, which includes the boost converters, charge pumps, battery chargers, buck converters, and linear regulators mentioned above. The function of power management circuits is to properly condition, convert, and deliver energy from various power sources to loads. Even though they do not directly provide users with a specific type of functionality or application, they are just as important because these blocks largely determine the energy efficiency of the whole system.

Though inefficient electronic systems have thousands of loop holes that lead to their inefficiency, an efficient one would always have an effective power management scheme with the supporting power management integrated circuit blocks. Discussions on system level power management scheme will be the topic of Chap. 2, and the design of power management IC will be explained in the next section.

1.4   Power Management IC Design

Integrated circuits (IC) today can be roughly classified into four major categories: analog, digital, radio frequency, and digital. Analog IC design deals with continuous time signal. Examples of analog IC blocks include amplifiers, oscillators, and comparators (Gray et al. 2001; Allen and Holberg 2002; Razavi 2001). Digital integrated circuits have been the driving force behind very-large scale integration (VLSI). Process scaling has allowed digital circuits to have more gates, faster speed, and lower power consumption. The introduction of hardware descriptive language such as Verilog and VHDL also fundamentally changed the design flow. Digital designers can now focus on logic with high level of abstraction, with strong EDA tools for synthesis and routing.

Radio-frequency (RF) ICs, also known as communication integrated circuits, feature unique circuitry such as low noise amplifier, oscillator, mixer, power amplifiers, and frequency synthesizer (Razavi 1997; Lee 2003). They are the only type of ICs that are extremely sensitive to high frequency effects and the parasitics of semiconductor processes. Mixed-signal integrated circuits refer to any ICs with both analog and digital contents, but they are more often used for analog-to-digital (ADC) and digital-to-analog (DAC) converters exclusively (Johns and Martin 1997).

The four categories are not clearly cut, and they can have substantial overlaps. Precise instrumentation amplifiers make use of a mixed-signal technique— chopper stabilization to overcome a fundamental limit in analog design, $1/f$ noise (Wu et al. 2009). Frequency synthesizers frequently use Delta-Sigma modulation, an extremely powerful digital technique, to improve close-in phase noise (Riley et al. 1993). Phase-locked loop (PLL), an essential RFIC block for clock and data recovery in transceivers, can be implemented in analog, mixed-signal, and all-digital fashions (Staszewski et al. 2005).

Previously unnoticed as scattered applications of analog IC blocks, power management integrated circuits recently emerge as a new and independent genre. Because of the need to improve power efficiency in either complex system-on-chips (SoC) or portable battery-powered devices, there is a surge of interest in power management circuits, including but not limited to low drop-out regulators (Rincon-Mora and Allen 1998; Leung and Mok 2003; Milliken et al. 2007; Al-Shyoukh et al. 2007), non-isolated switch-mode DC-DC converters (Lee and Mok 2004; Sahu and Rincon-Mora 2007; Xiao et al. 2004), single-inductor multiple-output converters (Ma et al. 2003; Le et al. 2007; Huang and Chen 2009), multi-phase DC-DC converters (Abedinpour et al. 2007; Hazucha et al. 2005), switched-capacitor regulators (Patounakis et al. 2004), and high-level integration challenges within complex power management integrated circuit (PMIC) chips (Shi et al. 2007).

This section will explore the characteristics and unique challenges associated with power management IC design. A close analogy with analog IC design is drawn, yet the interdisciplinary nature with power electronics is also revealed.

1.4.1   Challenges: High Efficiency and Full On-chip

The design considerations for power management circuits in a portable and low power green electronic system generally include:

- *Longevity.* A major limitation for many portable and battery-powered electronics is their short battery life, or the constant need for battery recharge and replacement. In applications involving large number of devices deployed at hard-to-reach locations, industrial hazardous environment, or military combat missions, this may be problematic, as the effort and expense to locate the devices and replace the batteries outweigh whatever benefits the electronics may offer.
- *Compactness.* Another requirement for portable applications is the compactness of the solution in all dimensions, including size, height, and weight. Large discrete components, such as transformers, could be too thick to fit in a slim smart phone. Too many discrete capacitors could take up considerable amount of board area, which would prevent the system from being used in small footprint science, medical, and military applications.
- *Manufacture-Ability.* In order for the whole green electronics system to be commercially feasible, the power management circuits, together with other circuit components, need to be manufacture-able within reasonable cost. In other words, all the electronic circuits within the system should be compatible with a lowest cost, mainstream semiconductor process so that all blocks can be integrated on the same chip or within the same package.

As a result, the ideal solutions that address those concerns should possess the following characteristics:

- *High Efficiency.* The meaning of efficiency is twofold: self-efficient and system-efficient. First, all the power converters, such as buck, boost, linear regulators, have to enjoy high efficiency themselves. For instance, low drop-out (LDO) topologies for linear regulators and synchronous rectifying structure for buck converter are preferred over high drop-out and free-wheeling topologies. Secondly, power management circuits need to be designed so that the system efficiency can be maximized. Therefore, system power consumption and budget has to be studied, and power converters design techniques that enable them to operate in various system power modes need to be analyzed.
- *Full On-chip.* Compactness at board level translates to both fewer off-chip components and smaller on-chip die area for integrated circuits. Off-chip capacitors, resistors, inductors, and transformers, should be kept at a minimum, and full on-chip power management IC topologies are highly desirable (Milliken et al. 2007). Meanwhile, on-chip die area efficiency is also important, as smaller die size would allow usage of smaller packages.
- *CMOS.* Although bipolar and BiCMOS technologies have been favorites for standalone power management integrated circuits, full CMOS design would be more desirable due to the possibility of higher-level integration with analog, digital, and RF circuitry.

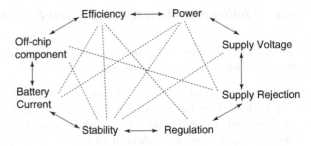

Fig. 1.5 Power management
IC design octagon

1.4.2 Specifications Versus Performance

This book seeks to make a distinction between design specifications ("specs")
and circuit performance. Specs refer to the conditions under which an electronic
circuit should operate, such as minimum and maximum output power, minimum
and maximum power supply voltages. Performance parameters, on the other hand,
describe how well the electronic circuit perform certain functions, such as efficiency,
slew rate, and power supply rejection.

In the real world of electronic product development, the two categories often
overlap. A datasheet of a product would often guarantee certain performance for
customers, making those performance parameters part of the "specs" during the
engineering development. Different "specs", on the other hand, can significantly
alter the ease of achieving many, if not all of the performance parameters, making
the comparison of any performance parameters meaningful only if the electronics
circuits are bound to the same "specs". Thus, this book will stay away from
associating any "specs" when discussing general circuit topologies, in the belief
that the readers, especially IC design engineers and product managers in industry
will be able to apply the design principles and trade-offs to different product and
application scenarios, achieving the maximum performance within given "specs".

1.4.3 Multi-Dimensional Trade-Off

Until very recently, discussions of power management IC blocks are scattered,
mostly in the context of various applications of operational amplifiers and feedback
(Gray et al. 2001). In a way, linear power management circuits have analog IC design at its core with additional power supply features. Non-linear power converters
are similar to some mixed-signal IC design, as they both deal with a significant
amount of clocking and switching. Similar to analog IC design (Razavi 2001), power
management integrated circuit design involves multi-dimensional trade-off.

Figure 1.5 shows the "power management octagon" as a counter part to the
"analog design octagon" (Razavi 2001). In addition to the common concerns of
voltage supply, power, noise, and stability, power management circuits are uniquely

challenged in their achievable efficiency. Output ripple (in switching converters) and ripple rejection (in linear regulators) are also essential, as power management circuits are primarily responsible for supply conditioning for other application circuits on chip. Finally, in the context of portable and battery-powered green electronics, off-chip components and battery current become important, as large on-chip components reduce the system portability, and large battery current reduces battery runtime.

1.4.4   Interdisciplinary with Power Electronics

It is only fair to mention that the study of power converters, i.e. electrical circuits that extract electrical energy from a source and deliver it to a load, converting the form of energy from one to another, is nothing new. They constitute the bulky of knowledge in the field of power electronics, which enabled some great engineering endeavor in human history, from the creation of electric power plants to laying down the nationwide power grid. Therefore, this book will not make an effort to present or re-state every power electronics principle that come along the way. Readers will benefit greatly from other books on these topics (Mohan et al. 2003; Erickson and Maksimović 2001).

Power management IC design is interdisciplinary, as it shares many of the power electronics principles, whether it is the steady-state voltage-second principle in the analyses of buck, boost, buck-boost, flyback, forward, half-bridge, or full-bridge topologies, or the Ampere law ($\oint \mathbf{H} \cdot dl = \sum I$) and the Faraday's law ($\varepsilon = -N\frac{d\Phi}{dt}$) regarding the interaction of electric and magnetic fields and components. In a sense, the inductive components and magnetic circuit analysis tell power management IC design apart from other IC genres: power management IC uses inductor as an energy storage element, as

$$V_L = L\frac{di_L}{dt} \tag{1.1}$$

$$e_L = \int_0^I Li\,di = \frac{1}{2}LI^2 \tag{1.2}$$

RFIC designs mostly use inductors in LC or RLC tanks for frequency selection, resonance, or impedance matching:

$$f_{L,C} = \frac{1}{\sqrt{2\pi LC}} \tag{1.3}$$

$$Z_L(\omega) = j\omega L \tag{1.4}$$

whereas analog, digital, and mixed-signal ICs barely uses any inductive components at all.

On the other hand, power management IC design is ultimately IC design, which leverages semiconductor and VLSI techniques. The large reservoir of analog, digital, and mixed-signal building blocks can be readily available when building portions of an integrated power converters, such as operational amplifiers, analog switches, gate drivers, and even delta-sigma modulators and PWM controllers. The disadvantages, however, include the voltage and power limitations: most semiconductor processes cannot render transistors with breakdown voltages V_{dss} above a few hundred voltages. The large current ratings in power converters also poses long-term reliability concerns in silicon and metals. Thus, electro migration rules for any specific process should be strictly followed.

1.5   Summary

Both the energy and environmental crises today point to improving energy efficiency as the most effective short term solution. Green electronics, in the sense of energy efficient electronics design, is the authors' answer, not only to the grand challenge of sustainable practices, but also to the on-going power challenge in consumer handsets, where longer battery life is demanded without sacrificing performance and experience.

In this chapter, a green electronics micro system prototype is envisioned, which incorporates communication, computing, control, sensing, and signal processing capabilities based on the needs from various applications. The power management section of the system is identified as the decisive portion when it comes to achieving higher efficiency, as it is responsible for energy delivery and implementing different power management schemes when necessary.

Finally, a quick look was taken at power management IC design compared to other major IC design genres, including analog, digital, RF, and mixed-signal circuits. Similarities were drawn between power management and analog IC design, yet unique challenges and the interdisciplinary nature were revealed as VLSI techniques merge with power electronics principles.

References

Zeb (2004) The zebranet wildlife tracker. URL http://www.princeton.edu/~mrm/zebranet.html
OMA (2008) OMAP3430 Multimedia Applications Processor. Texas Instruments, URL http://focus.ti.com/pdfs/wtbu/ti_omap3430.pdf
TIC (2009) A USB enabled system-on-chip solution for 2.4-GHz IEEE 802.15.4 and Zigbee Applications. Texas Instruments, URL http://focus.ti.com/lit/ds/symlink/cc2531.pdf
MC1 (2010) Advanced Zigbee-compliant Platform-in-Package (PiP) for the 2.4 GHz IEEE 802.15.4 Standard. Freescale Semiconductor, URL http://www.freescale.com/files/rf_if/doc/data_sheet/MC1322x.pdf

Vit (2010) Global environment outlook: environment for development (geo-4). URL www.unep. org

EPA (2010) Green electronics. URL http://www.epa.gov/oaintrnt/practices/electronics.htm

STM (2010) High-performance, IEEE 802.15.4 wireless system-on-chip. STMicroelectronics, URL http://www.st.com/stonline/products/literature/ds/16252/stm32w108cb.pdf

NRE (2010) National renewable energy laboratory. URL www.nrel.gov

EER (2010) The office of energy efficiency and renewable energy. URL www.eere.energy.gov

TPS (2010) Power Management IC for Li-Ion Powered Systems. Texas Instruments, URL http://focus.ti.com/lit/ds/symlink/tps65023.pdf

Ene (2010) U.s. epa energy star program. URL http://www.energystar.gov/

Abedinpour S, Bakkaloglu B, Kiaei S (2007) A multistage interleaved synchronous buck converter with integrated output filter in 0.18 um SiGe process. IEEE Trans Power Electron 22(6): 2164–2175, DOI 10.1109/TPEL.2007.909288

Al-Shyoukh M, Lee H, Perez R (2007) A transient-enhanced low-quiescent current low-dropout regulator with buffer impedance attenuation. IEEE J Solid-State Circuits 42(8):1732–1742, DOI 10.1109/JSSC.2007.900281

Allen PE, Holberg DR (2002) CMOS Analog Circuit Design, 2nd edn. Oxford University Press, USA

Carlson E, Strunz K, Otis B (2010) A 20 mv input boost converter with efficient digital control for thermoelectric energy harvesting. IEEE J Solid-State Circuits 45(4):741–750, DOI 10.1109/JSSC.2010.2042251

Dayal R, Dwari S, Parsa L (2010) Design and implementation of a direct ac-dc boost converter for low voltage energy harvesting. IEEE Trans Ind Electron PP(99):1–1, DOI 10.1109/TIE.2010. 2069074

Erickson R, Maksimović D (2001) Fundamentals of power electronics, 2nd edn. Springer Netherlands

Filani D, He J, Gao S, Rajappa M, Kumar A, Shah P, Nagappan R (2008) Dynamic data center power management: Trends, issues, and solutions. Intel Technology Journal 12(1):59–68

Gray PR, Hurst PJ, Lewis SH, Meyer RG (2001) Analysis and Design of Analog Integrated Circuits, 4th edn. John Wiley & Sons, Inc.

Guilar N, Kleeburg T, Chen A, Yankelevich D, Amirtharajah R (2009) Integrated solar energy harvesting and storage. IEEE Trans VLSI Syst 17(5):627–637, DOI 10.1109/TVLSI.2008. 2006792

Hazucha P, Schrom G, Hahn J, Bloechel B, Hack P, Dermer G, Narendra S, Gardner D, Karnik T, De V, Borkar S (2005) A 233-mhz 80%-87% efficient four-phase dc-dc converter utilizing air-core inductors on package. IEEE J Solid-State Circuits 40(4):838–845, DOI 10.1109/JSSC. 2004.842837

Huang MH, Chen KH (2009) Single-inductor multi-output (simo) dc-dc converters with high light-load efficiency and minimized cross-regulation for portable devices. IEEE J Solid-State Circuits 44(4):1099–1111, DOI 10.1109/JSSC.2009.2014726

Ipakchi A, Albuyeh F (2009) Grid of the future. IEEE Power Energy Mag 7(2):52–62, DOI 10.1109/MPE.2008.931384

Johns DA, Martin K (1997) Analog Integrated Circuit Design, 1st edn. John Wiley & Sons, Inc.

Juang P, Oki H, Wang Y, Martonosi M, Peh L, Rubenstein D (2002) Energy-efficient computing for wildlife tracking: Design tradeoffs and early experiences with zebranet. In: Int. Conf. Architecture Support Programming Languages Operating Syst., DOI 10.1109/ISCE.2009. 5156963

Kim S, Rincon-Mora G (2009) Single-inductor dual-input dual-output buck-boost fuel-cell-li-ion charging dc-dc converter supply. In: IEEE ISSCC Dig. Tech. Papers, pp 444–445,445a, DOI 10.1109/ISSCC.2009.4977499

Le HP, Chae CS, Lee KC, Wang SW, Cho GH, Cho GH (2007) A single-inductor switching dc-dc converter with five outputs and ordered power-distributive control. IEEE J Solid-State Circuits 42(12):2706–2714, DOI 10.1109/JSSC.2007.908767

Le T, Mayaram K, Fiez T (2008) Efficient far-field radio frequency energy harvesting for passively powered sensor networks. IEEE J Solid-State Circuits 43(5):1287–1302, DOI 10.1109/JSSC. 2008.920318

Lee CF, Mok P (2004) A monolithic current-mode cmos dc-dc converter with on-chip current-sensing technique. IEEE J Solid-State Circuits 39(1):3–14, DOI 10.1109/JSSC.2003.820870

Lee TH (2003) The Design of CMOS Radio-Frequency Integrated Circuits, 2nd edn. Cambridge University Press

Leung KN, Mok P (2003) A capacitor-free cmos low-dropout regulator with damping-factor-control frequency compensation. IEEE J Solid-State Circuits 38(10):1691–1702, DOI 10.1109/JSSC.2003.817256

Liserre M, Sauter T, Hung J (2010) Future energy systems: Integrating renewable energy sources into the smart power grid through industrial electronics. IEEE Ind Electron Mag 4(1):18–37, DOI 10.1109/MIE.2010.935861

Ma D, Ki WH, Tsui CY, Mok P (2003) Single-inductor multiple-output switching converters with time-multiplexing control in discontinuous conduction mode. IEEE J Solid-State Circuits 38(1):89–100, DOI 10.1109/JSSC.2002.806279

Martonosi M (2002) The wireless revolution. Tech. rep., Department of Electrical Engineering, Princeton University

Meehan A, Gao H, Lewandowski Z (2010) Energy harvesting with microbial fuel cell and power management system. IEEE Trans Power Electron PP(99):1–1, DOI 10.1109/TPEL.2010. 2054114

Milliken R, Silva-Martinez J, Sanchez-Sinencio E (2007) Full on-chip cmos low-dropout voltage regulator. IEEE Trans Circuits Syst I, Reg Papers 54(9):1879–1890, DOI 10.1109/TCSI.2007. 902615

Mohan N, Undeland TM, Robbins WP (2003) Power Electronics: Converters, Applications and Design, 3rd edn. John Wiley & Sons, Inc.

MOSIS (2011) On semiconductor c5 process. URL http://www.mosis.com/on_semi/c5/

Murugesan S (2008) Harnessing green it: Principles and practices. IT Professional 10(1):24–33, DOI 10.1109/MITP.2008.10

Ogunseitan O, Schoenung J, Saphores J, Shapiro A (2009) The electronics revolution: from e-wonderland to e-wasteland. Science 326(5953):670

Patounakis G, Li Y, Shepard K (2004) A fully integrated on-chip dc-dc conversion and power management system. IEEE J Solid-State Circuits 39(3):443–451, DOI 10.1109/JSSC.2003. 822773

Ramadass Y, Chandrakasan A (2010) An efficient piezoelectric energy harvesting interface circuit using a bias-flip rectifier and shared inductor. IEEE J Solid-State Circuits 45(1):189–204, DOI 10.1109/JSSC.2009.2034442

Razavi B (1997) RF Microelectronics, 1st edn. Prentice Hall

Razavi B (2001) Design of Analog CMOS Integrated Circuits, 1st edn. McGraw-Hill

Reiter U (2009) Perceived quality in consumer electronics - from quality of service to quality of experience. In: Proc. IEEE Int. Symp. Consumer Electron. (ISCE), pp 958–961, DOI 10.1109/ISCE.2009.5156963

Riley T, Copeland M, Kwasniewski T (1993) Delta-sigma modulation in fractional-n frequency synthesis. IEEE J Solid-State Circuits 28(5):553–559, DOI 10.1109/4.229400

Rincon-Mora G, Allen P (1998) A low-voltage, low quiescent current, low drop-out regulator. IEEE J Solid-State Circuits 33(1):36–44, DOI 10.1109/4.654935

Sahu B, Rincon-Mora G (2007) An accurate, low-voltage, cmos switching power supply with adaptive on-time pulse-frequency modulation (pfm) control. IEEE Trans Circuits Syst I, Reg Papers 54(2):312–321, DOI 10.1109/TCSI.2006.887472

Salerno DC, Jordan MG (2006) Methods and circuits for programmable automatic burst mode control using average output current

Saphores J, Nixon H (2007) California households' willingness to pay for greenelectronics. J Environmental Planning & Management 50(1):113–133

Schneiderman R (2010) Smart grid represents a potentially huge market for the electronics industry [special reports]. IEEE Sig Process Mag 27(5):8–15, DOI 10.1109/MSP.2010.937501

Shi C, Walker B, Zeisel E, Hu B, McAllister G (2007) A highly integrated power management ic for advanced mobile applications. IEEE J Solid-State Circuits 42(8):1723–1731, DOI 10.1109/JSSC.2007.900284

Shina SG (2008) Green Electronics Design and Manufacturing : Implementing Lead-Free and RoHS-Compliant Global Products. McGraw-Hill

Staszewski R, Wallberg J, Rezeq S, Hung CM, Eliezer O, Vemulapalli S, Fernando C, Maggio K, Staszewski R, Barton N, Lee MC, Cruise P, Entezari M, Muhammad K, Leipold D (2005) All-digital pll and transmitter for mobile phones. IEEE J Solid-State Circuits 40(12):2469–2482, DOI 10.1109/JSSC.2005.857417

Torres EOT, Rincon-Mora GA (2010) A 0.7- m bicmos electrostatic energy-harvesting system ic. IEEE J Solid-State Circuits 45(2):483–496, DOI 10.1109/JSSC.2009.2038431

de Vries RP (2010) Green chips: A new era for the semiconductor industry. In: Proc. IEEE 2010 Custom Integrated Circuits Conf., pp 1–2

Wu R, Makinwa K, Huijsing J (2009) A chopper current-feedback instrumentation amplifier with a 1 mhz 1/f noise corner and an ac-coupled ripple reduction loop. IEEE J Solid-State Circuits 44(12):3232–3243, DOI 10.1109/JSSC.2009.2032710

Xiao J, Peterchev A, Zhang J, Sanders S (2004) A 4-ua quiescent-current dual-mode digitally controlled buck converter IC for cellular phone applications. IEEE J Solid-State Circuits 39(12):2342–2348, DOI 10.1109/JSSC.2004.836353

Yoo J, Yan L, Lee S, Kim Y, Yoo HJ (2010) A 5.2 mw self-configured wearable body sensor network controller and a 12 uw wirelessly powered sensor for a continuous health monitoring system. IEEE J Solid-State Circuits 45(1):178–188, DOI 10.1109/JSSC.2009.2034440

Chapter 2
System Power Management

Power management for a portable and battery-powered product is ultimately a system level challenge. The total energy available from a battery is limited, and every component of the system will consume different amount of power, from the LCD touch screen to the advanced mobile digital signal processors. It may seem that the system level power management schemes go beyond an integrated circuit (IC) designer's concern. However, a clear understanding of these methods could give IC designers insights to the relative importance of different parameters within an IC block and effectively guide their transistor-level innovations.

This chapter first reviews some existing system level power management schemes, mostly notably a holistic approach consisting of system software, SoC architecture, and silicon IP (Carlson and Giolma 2008). Regarding low power, low duty-cycle portable electronics in particular, this chapter studies the relative importance of active mode and sleep mode, and concludes that sleep-mode efficiency is of greater importance, especially in highly integrated systems in advanced sub-micron processes. Potential battery life extension using a sleep-mode efficient DC-DC converter is estimated from reported experimental data and a log-linear prediction of leakage-supply dependence.

2.1 Overview: A Holistic Approach

A good example that embodies all the major challenges in portable power management is the design of a mobile smart phone. Consumers increasingly rely on their smart phones for communication, computing, and entertaining. Meanwhile, they want sleek, compact devices that can easily fit into their pockets. Integration at the chip level combining multiple processing cores in the same device, as well as the use of deep sub-micron to nanometer fabrication processes help reducing the size of wireless handsets, as the functions which used to require multiple silicon chips to implement can now be replaced with a single chip, saving significant board space. Unfortunately, deep sub-micron and nanometer processes exacerbate the problem

J. Hu and M. Ismail, *CMOS High Efficiency On-chip Power Management*, Analog
Circuits and Signal Processing, DOI 10.1007/978-1-4419-9526-1_2,
© Springer Science+Business Media, LLC 2011

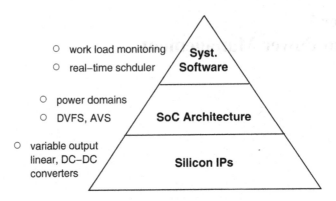

○   work load monitoring
○   real–time schduler
○   power domains
○   DVFS, AVS
○   variable output linear, DC–DC converters

Fig. 2.1 A holistic approach to system power management

of standby leakage power, and the added transistors and gates increase the dynamic power, making the task of battery extension more challenging (Carlson and Giolma 2008). It was estimated that an average of 10 times better power management scheme is needed from one technology node to another just to maintain the same level of talk and standby time, yet there is an increasing demand for longer battery despite all the technical difficulties (Royannez et al. 2005).

To address this challenge, some research and industrial products (Carlson and Giolma 2008; Royannez et al. 2005; Mair et al. 2007) have proposed or adopted a holistic approach to system power management, which include a bundle of technologies and intellectual properties (IP). Figure 2.1 shows a pyramid consisting of:

1. *System Software.* Various system power management software can run on mobile phone operating system (OS). An example of this can be a real-time task scheduler (Pillai and Shin 2001), which monitors the work load of the system, adaptively schedule tasks for execution while maintaining deadlines. With the help of task scheduler, power and performance can be boosted during heavy work load and critical tasks, while the opposite can be done during idle time to reduce consumption. These adjustments of power profiles can be either autonomous (Carlson and Giolma 2008), or via OS and user intervention.

2. *SoC Architecture.* In addition to software algorithm, certain SoC level architectural power management strategies can also be applied, such as power domain devision (Hattori et al. 2006), dynamic frequency and voltage scaling (DFVS) (Ma and Bondade 2010), adaptive voltage scaling (AVS) (Carlson and Giolma 2008), and memory retention (Narendra and Chandrakasen 2005; Wang et al. 2008).

   Power domain division divides a complex SoC cores into different power domains, each enjoying its own supply voltages and leakage current control (often through power gating). For example, mobile baseband processor OMAP 3,430 (OMA 2008) is divided into four to five power domains: MCU core, DSP core, graphic accelerator, Peripherals, and always-on circuits, each of which could be configured independently.

Another architectural power management scheme is dynamic voltage and frequency scaling (DVFS) (Ma and Bondade 2010). It is well known that the processing time T of a specific task and the dynamic power dissipated during computation can be modeled as

$$T = N/f \tag{2.1}$$

$$P_{dyn} = CV_{DD}^2 f \tag{2.2}$$

in which the frequency of operation is given by f and the total switching capacitance is C. For computation-intensive and latency-critical tasks such as video decompression, where short T is critical, raising the clock frequency f can expedite the execution of tasks. For low speed or idle tasks, only a fraction of throughput is required, and completing them ahead of their timing deadline leads to no discernible benefits. Reducing f for these tasks reduces the power consumption, but does not affect the total energy required to complete the task, which is $E = NCV_{DD}^2$. Another way of understanding is that even though power consumption is lower, the time required to execute the task becomes longer. Therefore, supply voltage V_{DD} needs to be reduced together with frequency f in order to significantly reduce the energy consumed (Burd et al. 2000). Used together with a real-time work monitor and scheduler , this technique can provide the optimum supply voltages and clock frequencies such that all computing tasks and processing needs are completed "just in time", thereby eliminating any slack period. A simple illustration of the principles of DVFS is shown in Fig. 2.2.

Adaptive Voltage Scaling (AVS) is similar to DVFS, but serves as a fine tune, self-adaptive power management scheme. It detects the variations in process, temperature and aging, and adaptively lowers or raises the voltage level accordingly (Mair et al. 2007). The hardware for AVS usually consists of some sensor circuitry, such as ring oscillators capable of self-monitoring the frequency of oscillation (or loop delay), which is often an indicator of device corners, temperature and aging, and the adjustable output power converters and oscillators, which can produce the various levels of output voltage and frequency required.

3. *Silicon IPs*. To support the architectural power management schemes, at least two types of silicon IPs are needed. One type is the various forms of integrated power converters that can execute the DVFS scheme by generating, distributing, and delivering the variable supply voltages to different circuits and power domains, such as multiple- and variable-output DC-DC converters (Ma and Bondade 2010; Huang and Chen 2009; Le et al. 2007; Ma et al. 2003). Due to the dynamic nature of DVFS, fast reference tracking DC-DC converters are also essential (Ma et al. 2004; Luo and Ma 2010). In addition, continual variation of output voltage is desirable, making switched-capacitor type converter less attractive (as they provide discrete supply voltages given by gain ratios (GR) (Ying et al. 2003)).

The other type of silicon IP needed is power configurable analog and digital sub-systems and building blocks, such as processors with multiple discrete

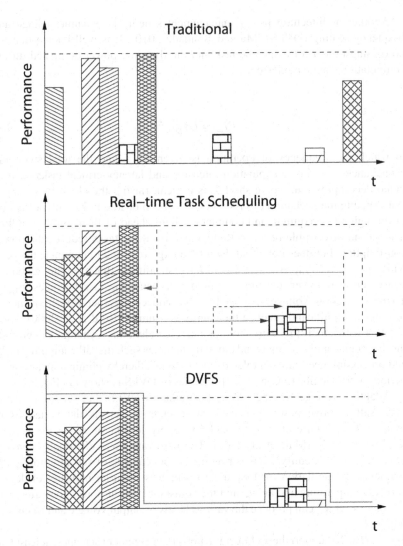

Fig. 2.2 DVFS versus traditional constant supply power management scheme

Operating and Performance Points (OPPs) for different application scenarios (full-speed, low-power, standby, and hibernation), multi-threshold (V_t) CMOS cells for different balances between performance and power consumption: low V_t MOSFETs to boost the performance during high-performance high-speed operation, and high V_t MOSFETs to suppress leakage during standby periods.

The study and development of power management software go beyond the scope of this book, but application-aware integrated power converters design is a key enabling factor within the holistic approach. Before a detailed discussion on that topic in the following chapters, however, the uniqueness of the system power management for very low power portable applications is first studied, revealing an interesting property of sleep-mode dominance, which will serve as a guiding motivation for the rest of the book.

2.2 Very Low Power Applications: A Sleep Mode Perspective

For portable and battery-powered applications, the success of system power management is often measured by battery runtime. Short battery life is not only inconvenient for consumer electronics like laptops and cell phones, but also fatal to the commercial success of very low power, multi-year electronic devices like IEEE 802.15.4 and Zigbee compliant wireless SoCs for industrial, scientific, and medical (ISM) applications.

Despite the innovations in battery technologies, from primary Ni-Cd and alkaline batteries to advanced rechargeable Li-Ion and Li-Polymer batteries, there will always be a limit on the total amount of energy available per size, weight, and reasonable cost. Consumers, however, are not only unwilling to sacrifice, but also craving for more functionality and better performance for newer portable and battery-powered electronics. Thus, increasing electronic circuits' efficiency remains the most effective way to prolong battery life.

The bottleneck of power efficiency improvement, however, lies in the sleep mode (Hu et al. 2010a). For example, battery-operated IEEE 802.15.4/Zigbee sensor nodes will be in sleep mode 99.9% of the time, waking up periodically for a few milliseconds to check a sensor or poll the other radios (Baumann 2006). Thus, the total power consumption will approach sleep mode power consumption, as depicted in Fig. 2.3, with duty cycle (wake up time per operating period) well below 10%.

However, reducing the current consumption of the chip, I_{cc}[1], during the sleep mode of a low power integrated circuit (IC) or System-on-Chip (SoC) is non-trivial. The sleep mode I_{cc} is usually made up of the biasing and quiescent current (I_q) of the key sub blocks that remains on during sleep, such as power-on-reset (POR), brown-out-detection (BoD), memory data retention, and sleep timer, as well as the leakage current of the rest of the circuits. It is challenging to reduce I_q of a circuit block without huge performance penalty, but it is even more challenging to control the leakage as the CMOS IC technology scales into the nanometer regimes, where leakage increases dramatically.

The reduction of standby leakage has been addressed by different techniques such as power gating (Singh et al. 2007), dynamic voltage scaling (DVS) (Burd et al. 2000), and body biasing (von Arnim et al. 2005). Power gating and DVS use the power supply voltage, V_{dd}, as the primary knob for reducing leakage currents,

---

[1] Notice that the total chip current (I_{cc}), more precisely the battery current (I_{batt}) decides the battery life, because the battery voltage remains relatively constant.

Current consumption

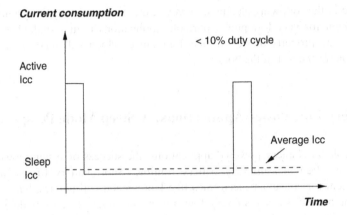

Fig. 2.3 Active and sleep mode current consumption for a IEEE 802.15.4/Zigbee compatible SoC

by either cutting off, or gating, a circuit from its power supply, or lowering V_{dd} to reduce leakage. The penalty is mostly the area and design overhead. Body biasing adjusts the threshold voltage, V_{th} to reduce transistor sub-threshold leakage. However, reverse body biasing worsens short channel effects like drain induced barrier lowering (DIBL), and increases V_{th} variation across a die, which makes it less effective with technology scaling (von Arnim et al. 2005).

Motivated by the important role the power supply, V_{dd}, plays in standby leakage reduction and sleep mode efficiency boosting, this section investigates the power chain structure within the widely used low power SoCs. A DC-DC converter based power chain is proposed to effectively reduce sleep mode battery current (Hu et al. 2010b).

2.2.1 Power Chain Study

A power chain can be loosely defined in this dissertation as a power flowing path from the battery to a point of load. Each load circuit will have one or more power chains to the battery. Depending on the power mode, different chains can be selected and activated.

Several examples of the power chains within in state-of-the-art commercial products are shown in Figs. 2.4, 2.5, and 2.6 respectively. Figure 2.4 shows a 2.4 GHz wireless SoC from STMicroelectronics built in ST's proprietary $0.18\,\mu m$ process (STM 2010). Circuit blocks such as General Purpose Input-Output (GPIO) and Power-on-Reset (POR) are directly powered from the battery due to voltage level compatibility and the need for direct battery voltage access. Analog, RF, and memory circuits are powered off an internal 1.8 V LDO, which provides the standard

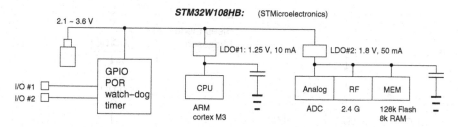

Fig. 2.4 Power chain example 1: STMicroelectronics STM32W108 2.4 G wireless SoC

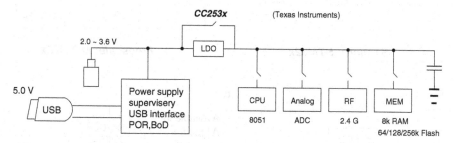

Fig. 2.5 Power chain example 2: TI CC253x 2.4 G SoC

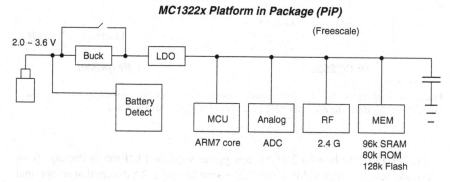

Fig. 2.6 Power chain example 3: Freescale MC1322x 2.4 G PiP

supply voltage of a $0.18\,\mu m$ process and allows for maximum IP reuse. The CPU, on the other hand, operates off a 1.25 V LDO to reduce power consumption.

Figure 2.5 shows a 2.4 GHz low power wireless SoC from Texas Instruments (TIC 2009). Unlike Fig. 2.4, both digital and analog sub-blocks are powered by a single 1.8 V LDO for simplicity. Notice that all sub-blocks can be disconnected from the supply rail during sleep mode to save on leakage, and the LDO can be by passed for direct battery operation.

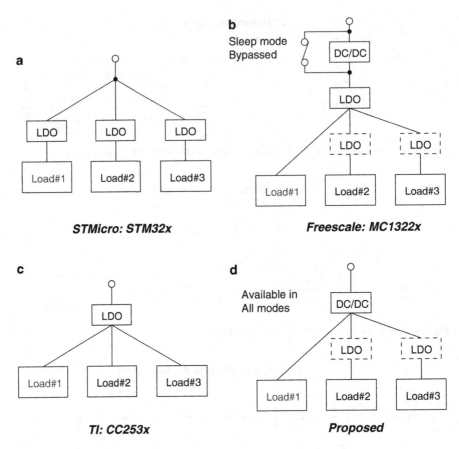

Fig. 2.7 Configuration of the sleep mode power chain: existing solutions and proposed. (**a**) *STMicro: STM32x* (**b**) *Freescale: MC1322x* (**c**) *TI: CC253x* (**d**) *Proposed*

Finally, Fig. 2.6 shows a 2.4 GHz low power wireless Platform in Package from Freescale Semiconductor (MC1 2010). It is similar to Fig. 2.5 except that an optional Buck converter is added in series of an LDO to create the internal 1.8 V power supply rail. According to its datasheet (MC1 2010), the converter will be activated during heavy load in active operations to save on battery current. During sleep mode, where the load current is low, the converter will be bypassed to avoid the Buck converter's operational overhead in power consumption.

In summary, the common theme for all three products is that the sleep mode power chain (the power chain to the load that remains on during sleep, e.g., Load 1 as in Fig. 2.7) consists of LDOs only, if not merely a wire. Even if a DC-DC converter is available in the structure, it is disabled during sleep mode. This section proposes using DC-DC converters in sleep mode so that the battery current in sleep mode will also be reduced.

2.2.2 Power Saving from Proposed Structure

This section will quantify the power saving by replacing the conventional LDO-based power chain with the proposed DC-DC-based structure during a leakage-dominated sleep mode operation for a hypothetical IEEE 802.15.4/Zigbee wireless SoC in a digital 65 nm CMOS process. Even though the analysis is based on a specific genre of low power wireless SoCs, the findings are nevertheless general and can be applied to other standby-power-critical applications as well.

The system is assumed to operate from a single-cell Li-Ion battery with nominal voltage of 3 V. During its sleep mode, $0.5\,\mu A$ of "always-on" static power consumption is assumed for the essential functional blocks operated directly off the battery supply. Leakage reduction techniques such as power gating are assumed to have been applied so that the leakage from RF and analog is negligible due to the small number of transistors involved. Leakage from the CPU is assumed to be less of an impact than that of the memory, since V_{dd} of combinational logic circuits, which usually make up the majority of CPU, can be set very close to zero without any major concern[2], but a V_{ddmin} is needed by a considerable portion of memory circuits like random-access memory (RAM) to retain their contents. To simplify the analysis, the sleep mode current budget will focus on the "always-on" and the leakage caused by data retention with V_{ddmin} of 500 mV.

The leakage current of a MOSFET is made up of gate tunneling, I_{gate}, gate induced junction leakage, I_{GIDL}, and the subthreshold current, I_{sub}, which often dominates and can be expressed as (Narendra and Chandrakasen (2005)):

$$I_{sub} = \frac{W}{L} I_t exp\left(\frac{V_{gs} - V_{th}}{nV_T}\right)\left[1 - exp\left(-\frac{V_{DS}}{V_T}\right)\right] \qquad (2.3)$$

As we can see from (2.3), the I_{sub} decrease exponentially as the supply voltage V_{DD} is reduced. Decreasing V_{DD} would also reduce I_{gate} and I_{GIDL} as well (Narendra and Chandrakasen 2005). To comprehensively model the leakage-supply relation, experimental data from a 65 nm ultra-low-power CMOS SRAM design (Wang et al. 2008) is used to construct an empirical formula.

As seen in Fig. 2.8 curve (a), the total leakage per bit decreases almost linearly with V_{DD} in the logarithm scale, which is in agreement with the exponential V_{DD}-relationship of I_{sub}, which is supposed to be dominant. Though additional techniques like PMOS back gate bias (b) and virtual ground raise (c) would reduce the leakage further as reported in Wang et al. (2008), this paper assumes a simple exponential I_{LEAK}-V_{DD} relation, i.e., assume

$$log(I_{LEAK}/bit) = k \cdot V_{DD} \qquad (2.4)$$

---

[2]In practice, sleep mode wake-up latency and current surge that leads to large IR drop is a concern (Juan et al. 2010).

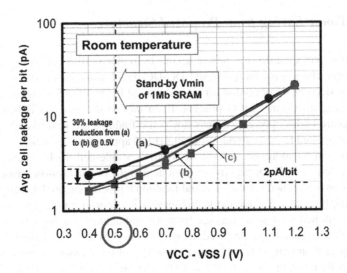

Fig. 2.8 Experimental data from 65 nm CMOS Ultra Low Power SRAM: Leakage reduction with V_{DD}

Given the data pair $(20pA/bit, 1.2V), (2.8pA/bit, 0.5V)$ in Wang et al. (2008), the parameter in 2.4 can be estimated as $\hat{k} = (k_1 + k_2)/2 = 2.38$. Assume that 16 KB of SRAM is on-board the SoC for data retention, the total leakage current can be modeled as

$$I_{LEAK} = 8bit/B \times 16KB \times exp(2.38 \cdot V_{DD})pA/bit \qquad (2.5)$$

The comparison is made between the conventional LDO based power chain and the proposed DC-DC converter, as seen in Fig. 2.9. The LDO is assumed to be ideal, thus

$$I_{BATT\,LDO} = I_{LEAK} + I_{Q\,LDO} \approx I_{LEAK} \qquad (2.6)$$

The DC/DC converter is assumed to have efficiency of η.

$$I_{BATT\,DCDC} = \frac{V_{DD}I_{LEAK}}{\eta \cdot V_{BATT}} \qquad (2.7)$$

Figure 2.10 shows the battery currents of the resulting system during sleep mode. A 30% efficient DC-DC converter would consume more battery current compared to the ideal LDO unless the RAM data retention voltage, V_{DD}, is lowered below 0.9 V. This is because an ideal LDO's efficiency improves as the drop-out voltage is reduced (This will be discussed further in Chap. 3), but DC-DC converter is assumed to have a flat 30% efficiency in this case. A 60% efficient DC-DC converter, on the

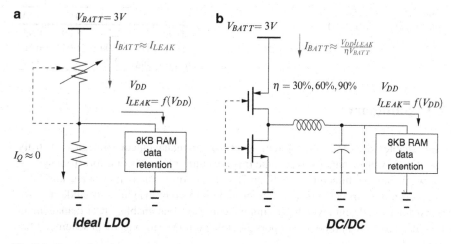

Fig. 2.9 Comparison between an LDO-based and a DC/DC based power chain in sleep mode. (a) *Ideal LDO* (b) *DC/DC*

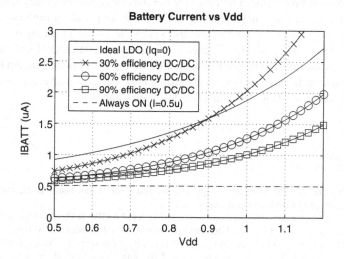

Fig. 2.10 Battery current reduction using DC/DC converter

other hand, would reduce the battery current by 35% at 0.9 V or 20–30% across the entire V_{DD} range of interest (0.5–1.2 V). A 90% efficient DC-DC converter would reduce the battery current even further. However, the physical design of a 90% efficient DC-DC converter silicon IP at light load is much more challenging (as will be discussed in Chap. 4). Table 2.1 summarizes the battery current savings from a 60% efficient DC-DC at three bench mark voltages.

Table 2.1 Battery current
(unit:μA) at different RAM
voltages (V_{DD})

V_{DD}	1.2 V	0.9 V	0.5 V
LDO	2.74	1.60	0.93
60% DC-DC	2.00	1.04	0.62
I_{BATT} saving	27%	35%	33%

2.3 Summary

This chapter discusses the system level power management strategies. A truly effective power management scheme involves multiple levels of effort from system software, SoC architecture to Silicon IPs. These discussions may seem unrelated to IC design at first glance, but they provide important insights and guidelines for transistor-level design, which is to innovate in ways that enable power management IC blocks to fully and better support the holistic power management scheme at the top level.

As an example of the said insight, the characteristics of very low power, low duty cycle battery-powered electronics are studied. Sleep mode efficiency is found to be the dominating factor in determining the overall system efficiency (compared to the active mode efficiency) because of the much longer sleep time the devices usually go through. Case studies on the system-level power chain of existing IEEE 802.15.4 Zigbee solutions reveals the lack of sleep-mode focus, and the battery runtime extension is estimated should different sleep-mode efficient DC-DC converter silicon IPs are available. This observation serves as a primary motivation for the transistor-level DC-DC converter design in Chap. 4.

References

OMA (2008) OMAP3430 Multimedia Applications Processor. Texas Instruments, URL http://focus.ti.com/pdfs/wtbu/ti_omap3430.pdf

TIC (2009) A USB enabled system-on-chip solution for 2.4-GHz IEEE 802.15.4 and Zigbee Applications. Texas Instruments, URL http://focus.ti.com/lit/ds/symlink/cc2531.pdf

MC1 (2010) Advanced Zigbee-compliant Platform-in-Package (PiP) for the 2.4 GHz IEEE 802.15.4 Standard. Freescale Semiconductor, URL http://www.freescale.com/files/rf_if/doc/data_sheet/MC1322x.pdf

STM (2010) High-performance, IEEE 802.15.4 wireless system-on-chip. STMicroelectronics, URL http://www.st.com/stonline/products/literature/ds/16252/stm32w108cb.pdf

NRE (2010) National renewable energy laboratory. URL www.nrel.gov

TPS (2010) Power Management IC for Li-Ion Powered Systems. Texas Instruments, URL http://focus.ti.com/lit/ds/symlink/tps65023.pdf

von Arnim K, Borinski E, Seegebrecht P, Fiedler H, Brederlow R, Thewes R, Berthold J, Pacha C (2005) Efficiency of body biasing in 90-nm CMOS for low-power digital circuits. IEEE J Solid-State Circuits 40(7):1549–1556, DOI 10.1109/JSSC.2005.847517

Baumann C (2006) The importance of sleep mode in zigbee/802.15.4 applications. Industrial Control DesignLine, URL http://www.industrialcontroldesignline.com/192701026

Burd T, Pering T, Stratakos A, Brodersen R (2000) A dynamic voltage scaled microprocessor system. IEEE J Solid-State Circuits 35(11):1571–1580, DOI 10.1109/4.881202

Carlson B, Giolma B (2008) Ti white paper: Smartreflex power and performance management technologies: reduced power consumption, optimized performance. URL http://focus.ti.com/pdfs/wtbu/smartreflex_whitepaper.pdf

Hattori T, Irita T, Ito M, Yamamoto E, Kato H, Sado G, Yamada Y, Nishiyama K, Yagi H, Koike T, Tsuchihashi Y, Higashida M, Asano H, Hayashibara I, Tatezawa K, Shimazaki Y, Morino N, Hirose K, Tamaki S, Yoshioka S, Tsuchihashi R, Arai N, Akiyama T, Ohno K (2006) A power management scheme controlling 20 power domains for a single-chip mobile processor. In: IEEE ISSCC Dig. Tech. Papers, pp 2210–2219, DOI 10.1109/ISSCC.2006.1696282

Hu J, Liu W, Ismail M (2010a) Sleep-mode ready, area efficient capacitor-free low-dropout regulator with input current-differencing. Analog Integr Circ Sig Process 63(1):107–112

Hu J, Liu W, Khalil W, Ismail M (2010b) Increasing sleep-mode efficiency by reducing battery current using a dc-dc converter. In: IEEE Int. Midwest Symp. Circuits Syst., pp 53–56

Huang MH, Chen KH (2009) Single-inductor multi-output (simo) dc-dc converters with high light-load efficiency and minimized cross-regulation for portable devices. IEEE J Solid-State Circuits 44(4):1099–1111, DOI 10.1109/JSSC.2009.2014726

Juan DC, Chen YT, Lee MC, Chang SC (2010) An efficient wake-up strategy considering spurious glitches phenomenon for power gating designs. IEEE Trans VLSI Syst 18(2):246–255, DOI 10.1109/TVLSI.2008.2010324

Le HP, Chae CS, Lee KC, Wang SW, Cho GH, Cho GH (2007) A single-inductor switching dc-dc converter with five outputs and ordered power-distributive control. IEEE J Solid-State Circuits 42(12):2706 –2714, DOI 10.1109/JSSC.2007.908767

Luo F, Ma D (2010) Design of digital tri-mode adaptive-output buck-boost power converter for power-efficient integrated systems. IEEE Trans Ind Electron 57(6):2151–2160, DOI 10.1109/TIE.2009.2034170

Ma D, Bondade R (2010) Enabling power-efficient dvfs operations on silicon. IEEE Circ Syst Mag 10(1):14 –30, DOI 10.1109/MCAS.2009.935693

Ma D, Ki WH, Tsui CY, Mok P (2003) Single-inductor multiple-output switching converters with time-multiplexing control in discontinuous conduction mode. IEEE J Solid-State Circuits 38(1):89–100, DOI 10.1109/JSSC.2002.806279

Ma D, Ki WH, Tsui CY (2004) An integrated one-cycle control buck converter with adaptive output and dual loops for output error correction. IEEE J Solid-State Circuits 39(1):140–149, DOI 10.1109/JSSC.2003.820844

Mair H, Wang A, Gammie G, Scott D, Royannez P, Gururajarao S, Chau M, Lagerquist R, Ho L, Basude M, Culp N, Sadate A, Wilson D, Dahan F, Song J, Carlson B, Ko U (2007) A 65-nm mobile multimedia applications processor with an adaptive power management scheme to compensate for variations. In: IEEE Symp. VLSI Circuits, pp 224–225, DOI 10.1109/VLSIC.2007.4342728

MOSIS (2011) On semiconductor c5 process. URL http://www.mosis.com/on_semi/c5/

Narendra S, Chandrakasen A (2005) Leakage in nanometer CMOS technologies. Springer

Pillai P, Shin KG (2001) Real-time dynamic voltage scaling for low-power embedded operating systems. In: Proc. ACM Symp. Operating Syst. Principles (SOSP), ACM, New York, NY, USA, pp 89–102, DOI http://doi.acm.org/10.1145/502034.502044

Royannez P, Mair H, Dahan F, Wagner M, Streeter M, Bouetel L, Blasquez J, Clasen H, Semino G, Dong J, Scott D, Pitts B, Raibaut C, Ko U (2005) 90nm low leakage SoC design techniques for wireless applications. In: IEEE ISSCC Dig. Tech. Papers, pp 138–589 Vol. 1, DOI 10.1109/ISSCC.2005.1493907

Salerno DC, Jordan MG (2006) Methods and circuits for programmable automatic burst mode control using average output current

Singh H, Agarwal K, Sylvester D, Nowka K (2007) Enhanced leakage reduction techniques using intermediate strength power gating. IEEE Trans VLSI Syst 15(11):1215–1224, DOI 10.1109/TVLSI.2007.904101

Wang Y, Ahn HJ, Bhattacharya U, Chen Z, Coan T, Hamzaoglu F, Hafez W, Jan CH, Kolar P, Kulkarni S, Lin JF, Ng YG, Post I, Wei L, Zhang Y, Zhang K, Bohr M (2008) A 1.1 GHz 12 uA/Mb-leakage SRAM design in 65 nm ultra-low-power CMOS technology with integrated leakage reduction for mobile applications. IEEE J Solid-State Circuits 43(1):172–179, DOI 10.1109/JSSC.2007.907996

Ying T, Ki WH, Chan M (2003) Area-efficient cmos charge pumps for lcd drivers. IEEE J Solid-State Circuits 38(10):1721–1725, DOI 10.1109/JSSC.2003.817596

Part II
Power Management IC Design

Chapter 3
Linear Regulators

3.1 Introduction

In the engineering world, it is often too easy and tempting to dive into the nitty-gritty details and lose track of the goal we originally set forth to accomplish. After all, the purpose of most power management circuits, regulators or converters, is to "power" certain loads (doesn't have to be circuits) and to "manage" the energy consumed smartly, especially in the context of green electronics. In a way, power management is the design of a more perfect power supply.

In this section, the power supply perspective is applied to analyze the need for both linear regulators and switching converters. Different control mechanisms to achieve a "smart" linear regulator are then studied. Finally, even within the same control mechanism, different circuit topologies are possible, and their advantages and disadvantages are also discussed.

3.1.1 A Power Supply Perspective

A general power supply can be modeled by the block diagram in Fig. 3.1. To meet the power demand from a load, a source of power is needed from which power can be extracted and processed. In addition, a smart power supply would also use the information from both the source and the load to adjust the power conversion accordingly, often for the sake of better performance and higher efficiency. Though power can take various forms in the real world, in this book we limit our discussions to electromagnetic power, which can be fully characterized by the vectors voltage ($\mathbf{V}$) and current ($\mathbf{I}$).

What constitutes a good power supply? For most electronic circuits and systems, a clean, stable, and accurate supply voltage is desired for smooth operations. Thus,

J. Hu and M. Ismail, *CMOS High Efficiency On-chip Power Management*, Analog Circuits and Signal Processing, DOI 10.1007/978-1-4419-9526-1_3,
© Springer Science+Business Media, LLC 2011

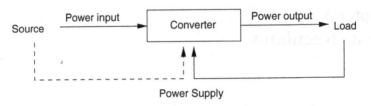

Fig. 3.1 Block diagram for a general power supply

most power supplies are designed to meet some or all of the following requirements (Mohan et al. 2003):

- *Conversion.* The source and the load may not have the same form of power. The power supply needs to convert whatever form, frequency, or value of input to match the need of the output. Four broad conversion categories can be identified: ac-to-ac, ac-to-dc, dc-to-ac, and dc-to-dc, and a power supply can involve one or multiple conversions.
- *Regulation.* The output voltage must be held constant within a specific tolerance for changes within a specified range in the power source and load.
- *Isolation.* The power source and load may be required to be electrically isolated completely, or partially within certain bandwidth, such that DC loading, high frequency noise, or other undesirable factors do not couple from the energy source to the load or vice versa.
- *Efficiency.* Ideally, all the power drained from the input should be delivered to the output. The power supply itself should dissipate as little energy as possible during the process.

In addition to these requirements, other factors such as the size and heat dissipation may also be important.

Given these requirements, what should a good power supply look like? An ideal voltage source will be a perfect candidate. As shown in Fig. 3.2a, an ideal voltage source provides perfect load regulation. No matter how the load impedance changes, the load voltage would always be stable. The output voltage would also be independent from its energy source, therefore bearing no influence from the input voltage.

Obviously, an ideal voltage source cannot be built in reality. It can only be approximated. Figure 3.2b shows a more elaborate model that includes some major nonidealities – a voltage controlled voltage source (VCVS). Though the output voltage is meant to be independent of the input, it remains a weak function of the input: $V_{out} = f(V_{in})$. As a result, noise and disturbance on the input source may pass through the power supply and corrupt the output signal. Meanwhile, there exists a finite output impedance R_S for the power supply. In some cases, it may even be a function of frequency and many other factors. Because of R_S, when load condition changes (e.g. reducing the resistive load), the output voltage shifts (an undershoot occurs). However, the closer the VCVS resembles an ideal voltage source, the better power supply it will be.

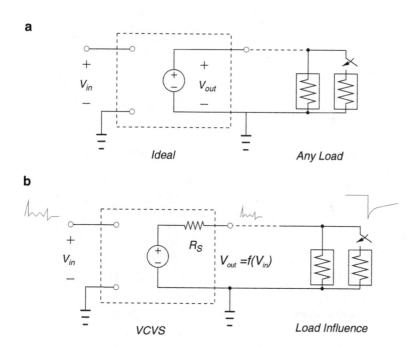

Fig. 3.2 Block diagram for a general power supply

Finally, among the four types of power conversion, this book focuses on DC-DC conversion. This is because most portable and wireless consumer applications power off a DC voltage source, whether it is a rechargeable battery or 5V USB voltage. The loads of interest are usually analog, digital, and mixed-signal integrated circuits that perform various functions. These circuits usually also require DC form of power supply. The differences between the input and output DC power lie not only in their absolute values, but most importantly in their accuracy, quality, and tolerance for variations. Hence the name linear and switching "regulators".

It is true that AC-DC converters are equally important in the presence of AC adapter. However, 50-60 Hz line input AC-DC converters usually divide the power conversion into two steps: 60 Hz AC to unregulated DC (*rectification*) and unregulated DC to regulated DC (*regulation*). Therefore, DC-DC regulation and conversion often prove more fundamental and they will be the primary topics of interest for the rest of the book.

3.1.2 *Feedback Mechanism*

From the previous section, a conclusion is drawn that building an optimum DC power supply is the same as designing a near ideal DC voltage source. Not only

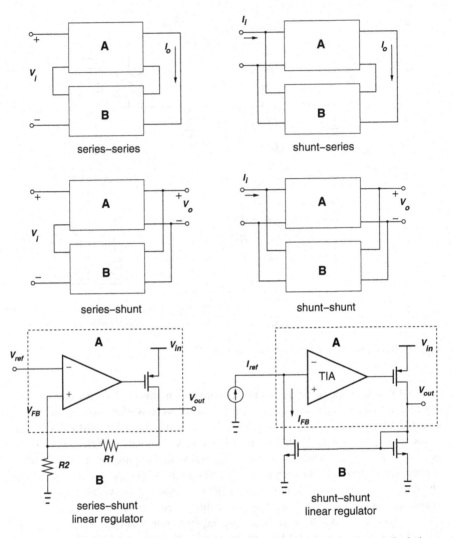

Fig. 3.3 Four major different feedback topologies and the suitable ones for power supply design

should the equivalent source impedance R_S be kept at a minimum, the input dependency should be as little as well, meaning that the f function in $V_{out} = f(V_{in})$ should be very weak.

Negative feedback is a widely used control mechanism to stabilize one or multiple parameters against variations in linear systems (Gray et al. 2001). The unique requirement of building an optimum voltage source dictates the type of feedback appropriate to use.

Assume that a feedback linear system can be divided into a basic amplifier (*A*) and a feedback network (*B*). Figure 3.3 shows the four basic feedback configurations: series-series, shunt-series, series-shunt, and shunt-shunt. Since R_S needs to be kept at a minimum, only series-shunt and shunt-shunt configurations should be considered, as they reduce the output impedance of the original network by the feedback loop gain.

$$R_o = \frac{R_S}{1 + AB} \tag{3.1}$$

The same conclusion can also be drawn by inspecting the essence of regulation. The output *voltage*, not current, is the parameter that needs to be stabilized against variations. Thus, only the output shunt feedbacks (series- or shunt-shunt) are preferred. Series-series and shunt-series feedback stabilize the output *current*, and therefore they are more suitable for building current sources.

Among the two shunt feedback configurations, series-shunt is more commonly used. Figure 3.3 shows a possible block level implementation of this configuration. An operational amplifier (opamp) and an MOSFET transistor constitute the basic amplifier A. A resistor ladder R_1, R_2 acts as the feedback network by sensing the output voltage V_{out} and feeding a portion of it back to the input (still in the form of voltage). This is a preferred configuration because opamps prefer voltage signals as input. The reference to be compared to can easily be a voltage reference, and often it can be bandgap reference circuits. The output voltage regulation is achieved because of the high gain of the opamp and MOSFET forces the two differential input signals to be equal, in other words

$$V_{out} \times \frac{R_2}{R_1 + R_2} = V_{ref}$$

$$\therefore V_{out} = \left(1 + \frac{R_2}{R_1}\right) V_{ref} \tag{3.2}$$

Shunt-shunt feedback is also a viable option. As shown in Fig. 3.3, if the output voltage is sensed, converted into current, and compared with a reference current, then a shunt-shunt linear regulator can be built. In this case, the basic amplifier will take current input differences and provide voltage output. This type of circuits are also known as *transimpedance amplifier*, or TIA for short. TIA are widely used in optical communication systems, where laser signals stimulate optical diodes to produce time varying current, which is further amplified.

Unlike series-shunt linear regulators, whose topologies are better known and widely used in industrial products, shunt-shunt linear regulators are rare. Some of the limiting factors for its usage include the extra effort involved in designing current-mode analog circuits. Voltage mode circuit IP, such as various opamps, are readily available. In addition, highly accurate current references are needed. If the current reference is to be derived from a voltage reference, than a buffer circuit is usually needed, as the commonly used bandgap reference does not have sufficient driving capabilities.

Though the analysis of feedback control so far has been in the context of linear systems, the same principles apply to non-linear systems as well. Figure 3.4 shows

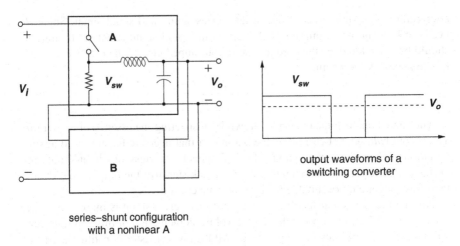

series–shunt configuration
with a nonlinear A

Fig. 3.4 Series-shunt feedback for a nonlinear system: switching converter

a series-shunt feedback used for a non-linear power supply. The major difference between a linear and a non-linear power supply is whether the input and the output are connected all of the time. In the linear regulator shown in Fig. 3.3, the power MOSFET always connects the power input to its output. In other words, the power MOSFET operates in Class A with 180° conduction angle.

In a non-linear power supply, however, the input power is connected to the output only part of the time. As seen in Fig. 3.4, the input is chopped by a switch and followed by an LC filter. This configuration is known as a buck converter. The conduction angle of the switch is less than 180°, and the output voltage (V_{out}) is proportional to the duty cycle of the chopper, i.e., the percentage of the time when the input power is connected.

Similar to its linear counterpart in Fig. 3.3, the switching converter in Fig. 3.4 can be broken into the basic amplifier (A) and the feedback network (B). The only difference here is that A is a non-linear circuit. As a result, the conclusion so far in this section and from Sect. 3.1.1 still applies: that an optimum switching DC-DC regular should also approximate an ideal voltage source with minimal R_S and input voltage dependence, and that output shunt feedback is preferred over series.

3.1.3 Topologies: LDO vs HDO

Even when implementing the most common feedback mechanism, a series-shunt linear regulator, the choice of the output stage configuration makes a significant difference in performance and power efficiency. This section will investigate the two major topologies: common-drain and common-source for the pass element, which lead to two distinctive classes: low drop-out (LDO)and high drop-out (HDO) of linear regulators.

Fig. 3.5 Series-shunt linear
regulator, whose controllable
impedance forms a voltage
divider with the load

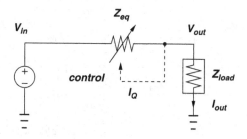

Before the study of topology, it is helpful to study the efficiency of a series-shunt
regulator. Instead of breaking down the regulator in A, B blocks, as in Fig. 3.3, it
can be viewed as an integrated power supply as shown in Fig. 3.5. Its impedance
Z_{eq} is controllable (via feedback), and it forms a voltage divider with the load
Z_{load}. Assume that the control of the source impedance Z_{eq} consumes no power,
the efficiency is independent of the load current I_{out}, and it is simply decided by the
input-output voltage (3.4).

$$V_{out} = \frac{Z_{load}}{Z_{load} + Z_{eq}} V_{in} \tag{3.3}$$

$$\eta = \frac{V_{out}}{V_{in}} \times 100\% \tag{3.4}$$

These reveal a basic limitation of a linear regulator: the drop-out voltage $V_{DO} =
V_{in} - V_{out}$.[1] The larger the drop-out voltage, the smaller the efficiency. In many
applications, the drop-out voltage is determined by system power management
requirement, such as the need to generate a 1.8 V supply rail from a 3 V I/O voltage,
or the need to regulate 1.2 V supply for low power digital circuits from 1.8 V internal
power rail.

A more detailed analysis of the linear regulator efficiency should also take into
account the output and quiescent current (I_{out} and I_Q). The controllable impedance
is also called the pass element (Lee 1999a), and it can be implemented using NPN,
PNP, NMOS, and PMOS (King 2000). Additional circuitry is needed to control the
pass element, and any current that is drain from V_{in} and does not flow to I_{out} is called
the quiescent current I_Q. Thus,

$$\eta = \frac{V_{out} \cdot I_{out}}{V_{in} \cdot I_{in}} = \frac{V_{out} \cdot I_{out}}{V_{in} \cdot (I_Q + I_{out})} \tag{3.5}$$

---

[1] Another definition sometimes used in industry and products is $V_{in} - V_{out}$ measured when V_{out} is
100 mV below the pre-fined output voltage, or in other words when the linear regulator is no longer
in regulation (Lee 1999a). The drop-out voltage defined in this manner will be $V_{DO,min}$. A higher
V_{DO} is always possible, but it further reduces η.

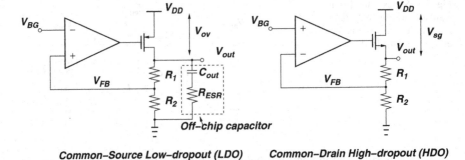

Fig. 3.6 Common-Source LDO versus Common-Drain HDO topology

As long as the quiescent current $I_Q \ll I_{out}$, the LDO efficiency can be approximated by the (3.4) very well. The ratio $\eta_i = \frac{I_{out}}{I_{out}+I_Q}$ remains important, and is often referred to as the current efficiency (Al-Shyoukh et al. 2007). In portable and battery-powered applications, it is important to keep the current efficiency high (e.g. above 90%) to prolong the system runtime.

Figure 3.6 shows two transistor-level implementations of Fig. 3.5. The left configuration uses a common-source PMOS as the pass element. Assume that M_1 operates in the saturation region, the minimum V_{DO} is approximately the overdrive of the pass element, typically below $200\,mV$. The configuration on the right, on the other hand, uses a common-drain (source-follower) NMOS pass element. The minimum V_{DO} in this case is $V_{gs,N} = V_{th,N} + V_{ov}$, which is at least one threshold voltage higher, assuming that the gate of M_1 cannot go above V_{in}. In this book, the two topologies are called low drop-out (LDO) and high drop-out (HDO) respectively.

From (3.4), it is obvious that an HDO is less efficient as an LDO. However, NMOS HDO has a number of performance advantages over PMOS LDO, regardless of its efficiency drawback.[2]

- *Stability*. NMOS pass element connected in a common-drain (source follower) configuration is a good buffer when driving capacitative loads. The output impedance of an HDO pass element is approximately $1/(g_{m,N} + g_{mb,N})$ instead of $r_{o,P}$ for an LDO, pushing the output pole of the linear regulator to very high frequency. Thus, the linear regulator would always be stable as long as the rest of the circuit (error amplifier) is properly compensated.
- *Area*. HDO linear regulators using NMOS pass elements require less die area, because a NMOS based pass element would be smaller than a PMOS one due to better mobility of majority carriers (electrons vs. holes): μ_n can be three times as large as that of the PMOS (holes) μ_p. Thus, to achieve the same $R_{ds,on}$, NMOS would need smaller W/L ratio, which translates to smaller area $W \cdot L$ given the

---

[2]Many of the parameters mentioned here will be discussed in details in the following sections.

same channel length L is used. This can be a major consideration for area efficient applications, because pass elements usually occupy most of the die area within a regulator.

- *Power Supply Rejection.* NMOS HDO would generally have better PSR because an NMOS common-drain output acts like a cascode device for the output node. PMOS common-source configuration, however, does little to shield the supply ripple, since the supply is directly connected to the source.[3]
- *Load transient.* NMOS HDO enjoys better AC line regulation due to better PSR. DC line regulation of a regulator depends on the DC open-loop gain, which is often sufficient in design ($\geq 60\,dB$) regardless of the pass element type. NMOS HDO is also superior in load regulation because of the source follower configuration, which has a smaller R_S ($1/(g_{m,N} + g_{mb,N})$) before and after close-loop feedback. Therefore, NMOS HDO regulator would experience less voltage undershot and overshot during abrupt load current changes.

Because of these advantages, researchers have studied methods to keep the NMOS HDO advantages while improving the efficiency, leading to topologies with NMOS pass element in a low drop-out (LDO) configuration, i.e. NMOS LDO. To do that, the circuit needs to generate a higher-than-supply gate voltage V_G. If $V_{G,max}$ can be made higher than V_{DD}, the NMOS drop-out voltage $V_{ds,N}$ would be as small as an overdrive, which is comparable to that of a PMOS LDO. A good example of NMOS LDO is den Besten and Nauta (1998). A charge pump is used to boost the NMOS gate voltage above supply. More recent examples include Kruiskamp and Beumer (2008), Camacho et al. (2009), Giustolisi et al. (2009), which use floating capacitor as voltage source. Their simplified block diagrams are shown in Fig. 3.7.

Higher-than supply V_G generation, however, becomes a major challenge for NMOS LDO . The high voltage switching involved in V_G generation (den Besten and Nauta 1998; Camacho et al. 2009; Giustolisi et al. 2009) introduces additional noise in the control loop. The noise attenuation relies solely on the loop gain at the switching frequency. Also, the circuit now needs additional capacitance, non-overlapping clock generation, and high voltage switches, in addition to the linear regulator core. Because of this, the design is more complicated, and the NMOS LDO would no longer have as significant amount of area saving as the pass element size suggested (μ_n/μ_p). Finally, higher-than-supply V_G may pose reliability threat, as the process feature size continues to scale down and the breakdown voltages of transistors are lower.

As a result, PMOS LDOs, though void of the advantages in stability, PSR, and load regulation, remains dominant in battery-powered applications because of its high efficiency, compactness if well designed, and simplicity (King 2000). Their performance has improved substantially thanks to numerous research on PMOS LDO design techniques. Unless other mentioned, the rest of the chapter will focus on PMOS LDO only.

---

[3]Nevertherless, PMOS LDOs can achieve very good PSR through common-mode ripple cancellation by using of PMOS mirror error amplifier (Gupta et al. 2004). The details are explained in section.

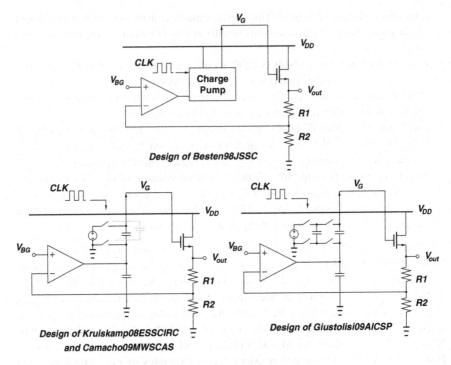

Fig. 3.7 NMOS LDO reported in literature

3.2 LDO Performance and Design Challenges

In this section, the commonly used LDO performance parameters and their design challenges are discussed. LDO performance parameters can be categorized into DC, AC and transient aspects. DC performance matrix includes line regulation, load regulation, drop-out voltage, quiescent current, and output accuracy. AC performance matrix includes stability, noise, and power supply rejection (PSR), and the transient performance includes line and load transient response, such as settling time, undershot and overshot.

3.2.1 DC Parameters

Like many other analog circuits, whose DC parameters mainly include the measure of gain, resolution, and accuracy, LDO DC parameters mostly describe the degree of accuracy in regulation. In addition, the quiescent current of the LDO with no load is also included.

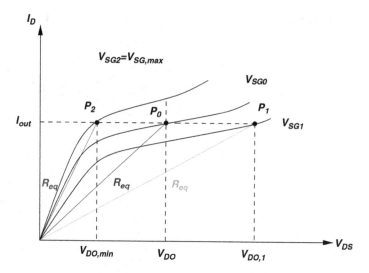

Fig. 3.8 Qualitative analysis of DC line regulation using PMOS I-V curve

3.2.1.1 DC Line Regulation

The DC line regulation measures the capability of the regulator to maintain V_{out} with varying V_{in} in a DC sense (Lee 1999b). Qualitatively, it can be analyzed using the MOSFET I-V curve in Fig. 3.8.

Suppose the regulator needs to source a constant I_{out} to the load and maintain a stable V_{out} while V_{in} changes slowly in a DC sense. The drop-out voltage $V_{DO} = V_{in} - V_{out}$, which also happens to be the drain-source voltage of the pass element, changes accordingly. If V_{in} increases above its nominal value, V_{DO} will increase to $V_{DO,1}$ in order to keep V_{out} constant. In the MOSFET I-V curve, this change is reflected by the shift of the operating point from P_0 to P_1, which resides on a curve with slightly lower V_{SG} value. This usually would not cause the circuit any trouble as long as V_{in} is not too high to cause drain junction of the pass element to breakdown. However, it does degrade the circuit efficiency, as suggested by (3.4).

If V_{in} decreases from its nominal value, V_{DO} will decrease, and the operating point will shift from P_0 to P_2. There is a limit, however, on how much V_{DO} can be reduced. Recall from Fig. 3.5 that the PMOS is a controllable impedance (resistance) in the DC sense. As the operating shift from P_0 to P_2, the controllable resistance R_{eq} is reduced accordingly. But the R_{eq} is ultimately lower bounded by $R_{ds,ON}$, which is the on-resistance of the MOSFET in deep triode region. In other words

$$R_{eq} \geq R_{ds,ON} = \frac{1}{\mu_p C_{ox} \frac{W}{L} (V_{SG} - V_{TH,p})} \tag{3.6}$$

Fig. 3.9 Quantitative line
regulation analysis

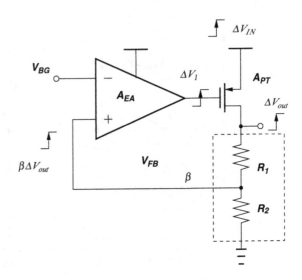

As V_{DO} is reduced, MOSFET will eventually enter the triode region because
the previous gain stage that is driving the pass element can only provide a limited
$V_{SG,max}$. As V_{DO} is further reduced, the pass element will reach the boundary of
saturation and linear region (depicted as P_2 in Fig. 3.8).

Quantitatively, the amount of DC line regulation can be defined and calculated
using Fig. 3.9.

$$\text{Line Regulation } = \frac{\Delta V_{out}}{\Delta V_{in}} \text{ or} \tag{3.7}$$

$$= \frac{\Delta V_{out}}{\Delta V_{in}} \cdot \frac{100}{V_{out}} \%/V \tag{3.8}$$

Let A_{EA}, A_{PT} be the DC gain of the error amplifier and the PMOS pass
element respectively. Suppose a DC voltage increase in the supply voltage ΔV_{IN}
is introduced, and the corresponding output voltage shift of ΔV_{OUT} follows:

$$\Delta V_1 = A_{EA} \cdot \beta \Delta V_{OUT} \tag{3.9}$$

$$\Delta V_{out} = A_{PT} \cdot (\Delta V_{IN} - \Delta V_1) \tag{3.10}$$

$$= A_{PT} \cdot (\Delta V_{IN} - A_{EA} \cdot \beta \Delta V_{OUT}) \tag{3.11}$$

$$\therefore \frac{\Delta V_{OUT}}{\Delta V_{IN}} = \frac{A_{PT}}{1 + \beta A_{PT} A_{EA}} \approx \frac{1}{\beta A_{EA}} \tag{3.12}$$

As a result, a higher error amplifier gain, which is part of the DC loop gain,
improves the line regulation. Notice that the derivation does not take into account

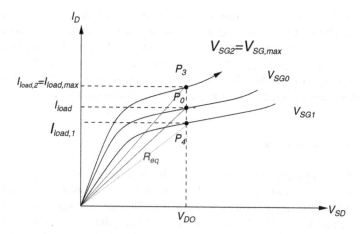

Fig. 3.10 Qualitative analysis of DC load regulation using PMOS I-V curve

factors such as the DC line regulations of the error amplifier or voltage reference, or the op-amp input offset voltages, all of which could introduce additional error.

In conclusion, DC line regulation is determined by the tunable range of R_{eq} of the pass element, given the output current and voltage it has to maintain. The method to improve line regulation includes up sizing the pass element (increasing W/L), and increasing the DC loop gain. However, the analysis in the following sections would reveal that these methods would lead to increased parasitics, slower slew rate, and instability. This gives the first example of the various trade-offs involved in power management IC design as indicated in Sect. 1.4.3. A common trade-off is to size the power MOSFET moderately such that at the maximum output $I_{out,max}$ and minimum input $V_{in,min}$, the MOSFET would still operate at the boundary of the saturation region (P_2).

3.2.1.2 DC Load Regulation

The DC load regulation measures the capability of the regulator to maintain V_{out} with varying I_{load} in a DC sense (Lee 1999b). Similar to line regulation, it can be analyzed through the MOSFET I-V curve in Fig. 3.10.

Suppose the regulator operates under a fixed V_{DO} to maintain a stable V_{out} while I_{load} changes slowly in a DC sense. I_{load} in this case happens to be the drain current I_D of the PMOS pass element. If I_{load} drops below its nominal value to $I_{load,1}$, V_{SG} will drop in order to source the same amount of I_{load}, and the operating point moves down from P_0 to P_4 accordingly.

If I_{load} increases from its nominal value, V_{SG} will need to increase as the operating point move up from P_0 to P_3. There is a limit, however, on how much V_{SG}

is available from the preceding driving stage. As the maximum V_{SG} curve is reached and I_{load} continues to rise, the operating point will have to move along the $V_{SG,max}$ curve to the right, which increases V_{DO} and causes the LDO to lose regulation. (V_{out} smaller than $V_{in} - V_{DO,defined}$). As a result, adequate DC load regulation requires a similar sizing strategy for the PMOS pass element, which is to guarantee its operation in saturation region at maximum I_{load} and minimum V_{DO}.

3.2.1.3 Drop-Out Voltage

As described above and in Sect. 3.1.3, V_{DO} directly affects the line regulation, load regulation, and achievable efficiency. Unfortunately, $V_{DO,min}$ is usually determined by the applications (input, output voltages and efficiency required) and beyond a circuit designer's control. However, if the drop-out voltage is flexible, it can be determined by the power transistor size, as the minimum V_{DO} for a MOSFET pass element is

$$V_{DO,min} = V_{DS,min} = V_{ov} = \sqrt{\frac{2I_D}{\mu C_{ox}(W/L)}} \tag{3.13}$$

The larger the pass element W/L is, the smaller $V_{DO,min}$ gets, and the higher η the LDO achieves.

3.2.1.4 Output Accuracy

The output accuracy of an LDO takes into account all the internal and external factors that may influence V_{out}, including line regulation, load regulation, reference fluctuation, temperature dependence, and transient response. As stated in Rincon-Mora (1996), let V_{out} and V_{ref} be the nominal output voltage and reference voltage, $V_{reference}$, the actual reference voltage applied at the feedback node, and β be the feedback factor.

$$V_{out,min} \leq V_{reference} \cdot 1/\beta + \Delta V_{load} + \Delta V_{line} + \Delta V_{TC} \leq V_{out,max} \tag{3.14}$$

where ΔV_{load}, ΔV_{line}, and ΔV_{TC} are voltage variations resulting from load regulation, line regulation, and temperature dependence respectively. The actual reference voltage $V_{reference}$ can be further expressed as

$$V_{reference} = V_{ref} + \Delta V_{TC,ref} + \Delta V_{linereg,ref} + \Delta V_{os} + V_{\varepsilon} \tag{3.15}$$

where V_{ref}, $\Delta V_{TC,ref}$, and $\Delta V_{linereg,ref}$ are the nominal reference voltage, reference temperature dependence introduced error, and reference line regulation introduced error respectively. Methods to reduce their effects are addressed through advanced voltage reference designs (Rincon-Mora 1996) which are beyond the scope of this

Fig. 3.11 Error voltage V_ε
due to finite DC loop gain

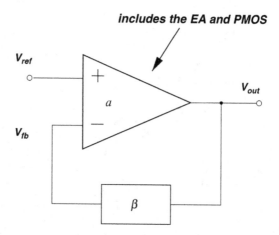

includes the EA and PMOS

V_{ref}

a

V_{fb}

V_{out}

β

book. ΔV_{os} is the input offset voltage of the error amplifier within the LDO. Methods of low offset designs are well documented in Gray et al. (2001) and Johns and Martin (1997), which usually involves eliminating the systematic offset (Johns and Martin 1997) and reducing random offset by larger geometry, smaller input overdrive, and better transistor matching (Gray et al. 2001).

Finally, there is also a term V_ε, which is the error introduced by limited DC gain of the LDO. This points to a fact that even if all the other circuit components are ideal, and all external error sources are not present, V_o would still be different from V_{ref}/β. As shown in Fig. 3.11, assume that the open-loop DC gain of the LDO is a and the feedback factor V_{fb}/V_o is β,

$$V_o = \frac{a}{1+a\beta} \cdot V_{ref} \qquad (3.16)$$

$$V_{fb} = \beta \cdot V_o = \frac{a\beta}{1+a\beta} \cdot V_{ref} \qquad (3.17)$$

$$\therefore V_\varepsilon = V_{ref} - V_{fb} = \frac{1}{1+a\beta} \cdot V_{ref} \qquad (3.18)$$

For a quick estimate of the magnitude of V_ε, in the LDO in Fig. 3.6, assume a 60 dB DC gain of the LDO, $R_1 = R_2$ (i.e. $\beta = 1/2$), $V_{BG} = 1.225\,V$, and $V_{out} = 2.45\,V$,

$$V_\varepsilon = \frac{1}{1+a\beta} \cdot V_{ref} = \frac{1}{1+10^3 \cdot 1/2} \cdot 1.225\ V = 2.445\,mV \qquad (3.19)$$

And the resulting output error from V_ε alone (other error sources ignored) can be found from (3.14) and (3.15):

$$\Delta V_{out} = \frac{\Delta V_{reference}}{\beta} = \frac{V_\varepsilon}{\beta} = \frac{2.445\,mV}{1/2} = 4.89\,mV \qquad (3.20)$$

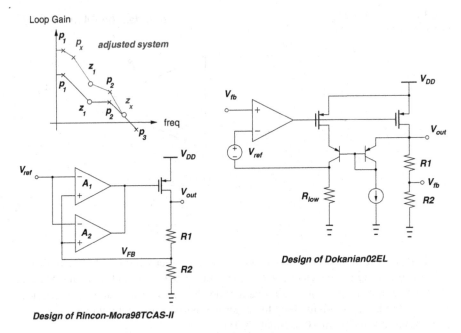

Fig. 3.12 Methods of improving DC parameters

In conclusion, when all other factors are equivalent, a higher DC open-loop gain would reduce ΔV_{load}, ΔV_{line}, and V_ε, which all help improve the LDO output accuracy. Unfortunately, increasing the open-loop gain alone would introduce a number of negative effectives to other performance parameters, such as stability. Methods in (Rincon-Mora and Allen 1998b) alleviated the trade-off by inserting a zero/pole pair, which increased the loop gain roll-off from 20 dB/dec to 40 dB/dec for most part of bandwidth, thus allowing higher DC loop gain and better load regulation. Another method to improve load regulation is to use floating V_{ref} (Dokania and Rincon-Mora 2002) to cancel load current change. A brief illustration of these methods are included in Fig. 3.12.

3.2.2 AC Parameters

Any linear circuits that take sinusoid or a combination of sinusoid inputs should be concerned with its frequency response. In other words, linear circuits usually have a limited amount of bandwidth within which the signal can be amplified and processed. In a DC-DC converter, however, both the input and output have "DC" type of signal. Any sinusoid signal that gets amplified and oscillates will void the purpose of the converter. As a result, a key AC parameter for an LDO is the ensemble of different stability measures.

Meanwhile, when sensitive analog and radio-frequency (RF) integrated circuits are used as loads of the LDO, any output ripples or noise could be detrimental to proper operation and corrupt the already vulnerable signals. Thus, AC parameters include power supply rejection as well. Though noise within a circuit is statistic in nature, it is treated as an AC parameter in this book because it is often frequency dependent.

3.2.2.1 Stability

The control of the pass element R_{eq} in Fig. 3.5 usually involves negative feedback, and the stability of the control loop should be guaranteed under all line and load conditions. Stability of a linear feedback system can be analyzed using either loop gain or return ratio (Gray et al. 2001). This book adopts the first approach, where the parasitic phase shift around a negative feedback loop is analyzed. When the parasitic phase shift reaches $180°$, the loop gain should be less than unity (0 dB); and when the loop gain drops to unity, the parasitic phase shift should be less than $180°$ with an appropriate amount of phase margin (Razavi 2001).

There are at least two poles in the LDO feedback loop. One is associated with the node of the error amplifier output (usually also the gate of the pass element) $p_{EA} = \frac{1}{R_{out,EA}C_{par}}$. The other is associated with the LDO output, $p_{out} = \frac{1}{R_{out}C_{out}}$. When both large output current and low quiescent current are required, as is often the case in portable consumer applications (Rincon-Mora and Allen 1998a; Al-Shyoukh et al. 2007), the pass element is sized larger to provide the extra output current capabilitiy without degrading $R_{ds,on}$, and the output impedance of the error amplifier is also increased, as channel lengths are often prolonged to accommodate low power operations. As a result, p_{EA} becomes substantially lower so that it becomes comparable to p_{out}, making the feedback loop potentially unstable with two closely located low frequency poles.

A popular way to compensate such an LDO involves introducing a zero within the feedback loop. In applications where large C_{out} were used, a series resistor can be inserted with C_{out}, or a minimum equivalent series resistance (ESR) of C_{out} is required such that a left-half plane (LHP) zero $z_{ESR} = \frac{1}{R_{ESR}C_{out}}$ can be created. The C_{out} and ESR value can be further specified to cancel p_{EA} if desired.

The downside of this compensation method is that the ESR values of the external capacitance need to be tightly controlled. First, the ESR value has to be above a minimum value. If the ESR is small, z_{ESR} is located at too high a frequency that it does little to improve the phase margin. If ESR alone is used to compensate an LDO, the required value could be well above 1 Ω (Rincon-Mora and Allen 1998a). This would rule out the use of any cheap and compact ceramic capacitance, which is essential for small footprint electronic systems. Secondly, even if a large ESR value is available, the resistance would introduce undesirable voltage spikes, especially when C_{out} charges and discharges itself during load transients. Last but not the least, the ESR of a capacitance shows dependency on temperature and voltage, among

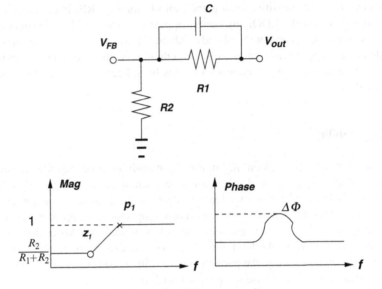

Fig. 3.13 Analysis of the feedback network with bypass capacitor

others. If the nominal location of z_{ESR} perfectly cancels p_{EA}, the exact location of z_{ESR} across different components, voltage, and temperature shifts significantly and leads to a zero-pole doublet, which is very undesirable for time-domain response (Gray et al. 2001; Razavi 2001).

As a result, methods that introduce a zero without relying on external ESR have been invented. A bypass capacitance can be inserted in parallel with the resistive feedback network, as seen in Fig. 3.13. Strictly speaking, this method introduces both a zero and a pole. But the zero is located before the pole such that a net phase boost can be reaped from the configuration. Quantitatively, the transfer function of the feedback network can be written as

$$V_{fb} = V_o \times \frac{Z}{Z + R_2} \tag{3.21}$$

$$\therefore \frac{V_{fb}}{V_o} = \frac{R_1 Cs + 1}{R_1 Cs + (1 + R_1/R_2)}$$

$$= \frac{R_2}{R_1 + R_2} \left(\frac{1 + sCR_1}{1 + sCR_e} \right) \quad \text{where } R_e = R_1 \| R_2 \tag{3.22}$$

The pole and zero can be found at

$$\omega_z = -\frac{1}{R_1 C} \tag{3.23}$$

$$\omega_p = -\frac{1 + R_1/R_2}{R_1 C} \tag{3.24}$$

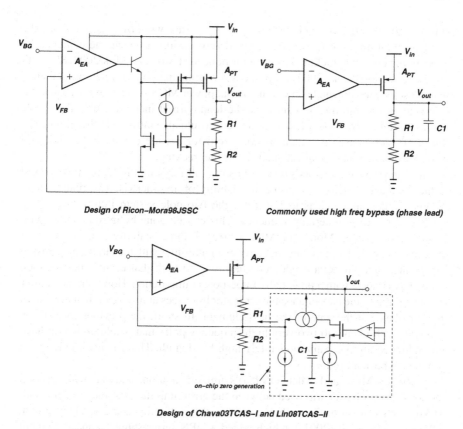

Fig. 3.14 Frequency compensation methods of a PMOS LDO

Since the zero comes before the pole but they are closely located, the phase boost from the structure is limited. The maximum phase boost ($\Delta\phi$) can be calculated as

$$\Delta\phi = 2\arctan\sqrt{\frac{\omega_p}{\omega_z}} - 90° \le 19° \tag{3.25}$$

Another method is to create an on-chip zero using active circuitry. A voltage controlled current source (VCCS) was proposed (Chava and Silva-Martinez 2004; Lin et al. 2008) to replace the bypass capacitance C in the phase-lead network above, as seen in Fig. 3.14. The idea is to eliminate the pole which was associated with the zero in the phase-lead network. Therefore, a nearly 90° phase boost is possible, instead of the limited amount as indicated in (3.25). Furthermore, cheaper and smaller multilayer ceramic capacitors can be used instead of the bulky tantalum capacitors, since their ESR no longer need to act as the main compensation factor.

In state-of-the-art LDO design, another major challenge in stability is that LDO needs to be source a wide range of output current. As a result, p_{out} changes substantially, as the pass element $r_{ds} = \frac{1}{\lambda I_D}$ varies with I_D (I_{out}), making the compensation more difficult. Since a fixed zero no longer meets the demand, the concept of zero-pole tracking has been introduced (Kwok and Mok 2002; Shi et al. 2007). Kwok and Mok (2002) proposed a zero-tracking pole compensation theme, which uses scaled downed copy of the output power MOS in series with a capacitor to create an on-chip tracking zero. As the load current increases, the output pole moves to higher frequency. The tracking zero would also move to higher frequency. Shi et al. (2007) used a similar tracking zero for a wide output range LDO and provided detailed mathematical analysis for the tracking.

The compensation methods discussed so far all rely on introducing phase boost in the feedback system to improve the stability of an otherwise two-pole system. Another major approach is to transform the two-pole system into a single pole system, at least up to certain frequency. This can be done by introducing a gate-drive buffer (Rincon-Mora and Allen 1998a). From a stability point of view, it disconnects the high output impedance of error amplifier with the large gate parasitic capacitance, and breaks a single node into two. At first glance, the feedback loop has one extra stage with a potentially three-pole configuration. However, the emitter follower reduces the capacitance the error amplifier sees at its output. It also reduces the impedance at the gate of the pass element. In short, the previous p_{EA} at low frequency is broken into two very high frequency poles that would no longer have any influence up to certain moderately high bandwidth. Thus, a single pole (p_{out}) system can be achieved.

In Rincon-Mora and Allen (1998a)'s implementation, a dynamically biased emitter-follower buffer was used between the error amplifier and the pass element. Al-Shyoukh et al. (2007) further improves the buffer by using a super source follower (Gray et al. (2001)), which added a NPN series-shunt feedback element to a PMOS source follower. The output impedance seen from the gate was further reduced by the current gain of the bipolar device (β). Even though an NPN was used as the feedback element in Al-Shyoukh et al. (2007), an NMOS would work equally well. A brief illustration of these methods are included in Fig. 3.14.

3.2.2.2 Power Supply Rejection

A major advantage of linear regulator is its power supply rejection (PSR), which shows the LDO's ability to reject AC power supply ripples at different frequencies. High power supply rejection ratio (PSRR) makes linear regulators suitable for post regulation after switching converters, whose efficiency is higher than linear regulators but suffers from large output ripples. PSR can be defined as $PSR = \frac{\Delta V_{out}}{\Delta V_{dd}}$. Though PSR of operational amplifiers are well documented in Steyaert and Sansen (1990) and Gray et al. (2001), PSR for linear regulators was not well studied until Gupta et al. (2004) qualitatively described the parameter as follows.

PSR of an LDO can be approximated by the superposition of the inverse of the loop gain within UGF and the output filtering beyond the UGF, as depicted

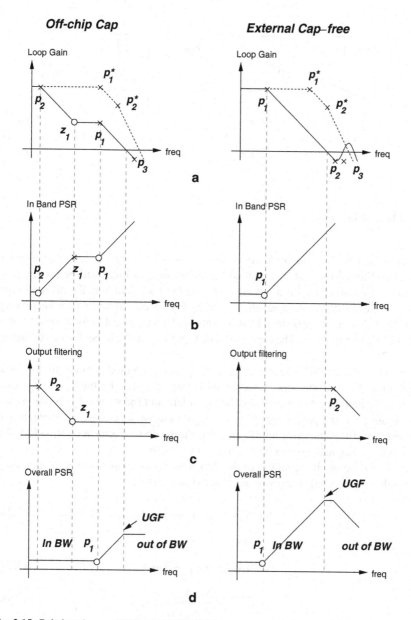

Fig. 3.15 Pole locations and PSR of a PMOS LDO

in Fig. 3.15: row (a) is the loop gain, row (b) is the inverse of (a), row (c) is the low-pass filtering due to C_{out} and $R_{out,LDO}$ (p_2), and row (d) is the superposition of row (b) and row (c). Notice that from row (a) to row (b), zeros and poles in loop gain transfer function become poles and zeros in the reverse of loop gain, denoted

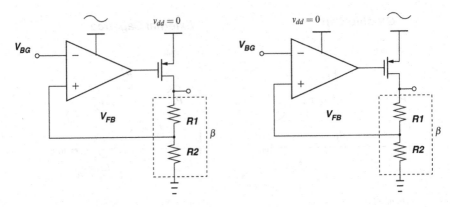

Fig. 3.16 PSR derivation

as "in-band PSR" in Fig. 3.15. Also notice that in row (c), the low-pass filtering would be limited by ESR, i.e. beyond frequency of z_{ESR} there will be no low-pass filtering effect since C_{out} is a virtual short and the PSR is merely the division ratio of $R_{ds,PMOS}$ and R_{ESR}. As a result, p_2 and z_1 cancel each other, which keeps very good PSR (low in $y-$ axis since PSR is expressed in negative dB) for a conventional PMOS LDO on the left. (The capacitor-free LDO scenario will be discussed later in Sect. 3.2.3.)

Quantitatively, PSR can be calculated using a method similar to DC line regulation. Assume the same notation of letting A_{EA}, A_{PT} be the open-loop gain of the error amplifier (1st stage) and the pass element (2nd stage), A_{p1} be the power gain (gain from the power supply to the signal output, as defined in Steyaert and Sansen (1990)) of the error amplifier, β be the feedback factor $R_2/(R_1 + R_2)$. An AC small-signal analysis can be drawn as in Fig. 3.16.

First, set the small-signal v_{dd} of the 2nd stage (pass element) to zero, assuming all ripples are injected from error amplifier supply rail.

$$\because v_{EA} = \beta \cdot v_{out} \cdot A_1 + v_{dd} \cdot A_{p1} \tag{3.26}$$

$$\therefore v_{out} = v_{EA} \cdot A_{PT}$$

$$= \beta A_{EA} A_{PT} v_{out} + A_{PT} A_{p1} v_{dd} \tag{3.27}$$

$$\therefore v_{out} = \frac{A_{PT} A_{p1}}{1 - \beta A_{EA} A_{PT}} v_{dd} \tag{3.28}$$

Next, set the small-signal v_{dd} of the 1st stage (error amplifier) to zero. All ripples are now injected from the source of the pass element.

$$\because v_{EA} = \beta \cdot v_{out} \cdot A_{EA} \tag{3.29}$$

$$\therefore v_{out} = (v_{EA} - v_{dd}) \cdot A_{PT}$$

$$= \beta A_{EA} A_{PT} v_{out} - A_{PT} v_{dd} \tag{3.30}$$

$$\therefore v_{out} = \frac{-A_{PT}}{1 - \beta A_{EA} A_{PT}} v_{dd} \tag{3.31}$$

Finally add the two together according to the property of superposition.

$$v_{out} = \frac{A_{PT} A_{p1}}{1 - \beta A_{EA} A_{PT}} v_{dd} + \frac{-A_{PT}}{1 - \beta A_{EA} A_{PT}} v_{dd} = \frac{(1 - A_{p1}) A_{PT}}{1 - \beta A_{EA} A_{PT}} v_{dd} \tag{3.32}$$

Notice that if $A_{p1} = 1$, i.e., the error amplifier has unity power supply gain (no power supply rejection), $v_{out} = 0$ regardless of the amplitude of v_{dd}. If $A_{p1} = 0$, $v_{out} = \frac{A_{PT}}{1 - \beta A_{EA} A_{PT}} v_{dd}$. As a result, $A_{p1} = 1$ turns out to be best for the PSR of a PMOS-based LDO. The counter-intuitive conclusion can be understood by the fact that $A_{p1} = 1$ means that an identical copy of the supply ripple being is fed to the pass element's gate. Consider that the same supply ripple is also present at the pass element's source, the net effect of any ripple is canceled ($V_{SG,p} = V_S - V_G$).

Following this principle, a number of PSR enhancing techniques have been proposed. A resistive divider stage is often inserted between the pass element and error amplifier gain stage, either with low impedance to the power supply (diode-connected MOSFETs, Hoon et al. (2005)) or high impedance to the ground (cascode device, Wong and Evans (2006)). As a result, the supply gain A_{p1} was significantly increased and PSR of the LDO greatly improved. El-Nozahi et al. (2010) proposed a ripple feed forward path that tracks V_{gate} ripple with respect to V_S ripple in more accurate fashion. Other approaches to boost PSR include cascoding the pass element (Ingino and von Kaenel 2001; Wong and Evans 2006; Gupta and Rincon-Mora 2007). Cascoding effectively increases the output impedance of pass element, which helps improve out-of-band PSR at high frequency, which is ultimately limited by the voltage division from the supply to the output node. However, V_{DO} increases with cascoding, and that degrades the regulator efficiency. A brief illustration of these methods are included in Fig. 3.17.

3.2.2.3 Noise

When an LDO is used to generate power supply to RF blocks such as VCO, output noise of the regulator becomes a major concern. Typically, noise in an LDO is specified as output-referred noise (Teel 2005). The input referred noise of the LDO is first found, then multiplied by the close-loop gain, V_{out}/V_{BG}.

The major contributors to LDO output noise are the bandgap reference, feedback resistors, and the error amplifier input stages. Interestingly, the large pass element does not contribute a significant amount of output noise due to the large g_m, which reduces its noise contribution when referred to the gate input (Wang and Bakkaloglu 2008).

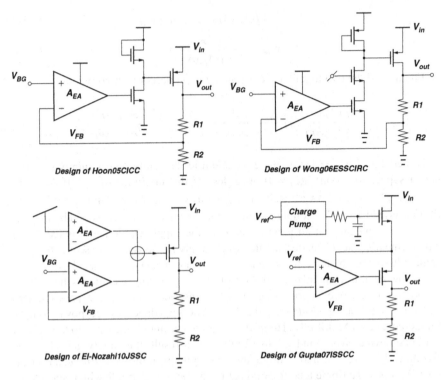

Fig. 3.17 Power supply rejection enhancement methods for a PMOS LDO

A low noise voltage reference is needed for a low noise LDO design. A common method is to use a large RC filter. However, it is difficult to filter out the $1/f$ noise, and a large capacitor could slow down the regulator start-up. To reduce the thermal noise from the feedback resistor would require using smaller resistors, which leads to more quiescent current flowing through the pass element and the feedback network. Finally, the error amplifier input stage contributes to the input-referred noise (hence output noise, as they are related by a constant V_{out}/V_{BG} factor). $1/f$ flicker noise is usually dominant, and it can be reduced by increasing W and L of the input pair. The penalty is obviously a larger area. Similar to low noise instrument amplifier, chopper stabilization techniques (Oh et al. 2008) can be applied to the error amplifier and reduce $1/f$ noise accordingly.

3.2.3 Transient Parameters

As a power supply module, LDO would face various degree of load variations, such as sudden load increase and decrease from digital processors as their computing need changes. As a result, the output voltage of the LDO would fluctuate, experience

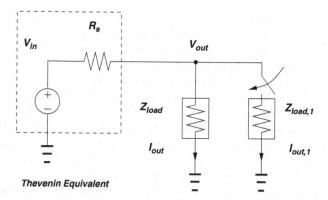

Fig. 3.18 Thevenin equivalent circuit in analyzing overshot and undershot

undershot and overshot, and take finite amount of time to settle to the new values. If we apply the power supply perspective in analyzing the problem, as explained in Sect. 3.1.1, the transient glitches are ultimately results of a non-ideal power supply. If $R_S = 0$, then there will be no undershot or overshot whatsoever.

The undershot and overshot problem can also be qualitatively analyzed using Thevenin equivalent circuit shown in Fig. 3.18.

Assume that at time t_{0-} that a load branch $I_{out,1}$ is switched on. The overall load resistance suddenly decreases. In the small-signal sense, if the Thevenin equivalent of linear regular has a high internal impedance, then the output voltage would change dramatically from t_{0-} to t_{0+}, which will cause a huge undershot. The magnitude of the undershot would be related to the small-signal impedance at the output node of the linear regulator, as well as the loop response time. As a result, transient response can be improved by either increasing the circuit bandwidth, or reducing the small-signal output impedance.

In the case of an LDO, the finite response time is due to the finite loop bandwidth (BW) and Error Amplifier (EA) Slew Rate (SR) (Rincon-Mora and Allen 1998a):

$$\Delta t_1 \approx \frac{1}{BW} + t_{SR} = \frac{1}{BW} + C_{par}\frac{\Delta V_{out,EA}}{I_{SR}} \tag{3.33}$$

Based on (3.33), different efforts can be made to boost the transient performance. For example, increasing I_Q of the error amplifier would increase the bandwidth and improve the slew rate at the pass element gate, as $SR \propto I_Q/C_{par}$. The obvious downside is the increase in power consumption.

The bipolar emitter-follower buffer proposed in Rincon-Mora and Allen (1998a) not only helped with stability (as discussed in Sect. 3.2.2.1), but also improve the load transient response without major increase in I_Q. This is because the error amplifier sees a much smaller C_{be} instead of a large C_{gate} at its output, thus it can be designed to be more power efficient for the same bandwidth requirement, or improve the bandwidth given the same current budget. In addition, the buffer is dynamically biased. When load current increases, more current is mirrored from the output into the buffer, increasing the gate drive current I_{SR} during slewing periods. Notice that the dynamic biasing does not increase the quiescent current of the circuit. The extra buffer current is only copied when the output current increases substantially. Al-Shyoukh et al. (2007) further improved the buffer through Buffer-Impedance Attenuation (BIA) technique. An NPN bipolar transistor was used in series-shunt negative feedback, a configuration otherwise known as a super source-follower (Gray et al. 2001). This structure reduces the buffer output impedance by approximately the current gain (β) of the bipolar transistor.

Transient response can also be improved by reducing the small-signal impedance at the output node. Oh and Bakkaloglu (2007) proposed using a current-feedback amplifier (CFA) as the buffer between the error amplifier and the pass element. The CFA then introduces a low ac impedance path at the output, which achieves fast transient with low power consumption. Hazucha et al. (2005) used dual-loop replica biasing, with a low BW loop for voltage regulation and a high BW low-dropout source follower for fast transient. It also leverages voltage positioning so that $\Delta V_{AC} = \Delta V_{DC}$, or in this book's notation, letting R_S in Fig. 3.18 be equal to R_{eq} in Fig. 3.5. Man et al. (2007) replaced the conventional error amplifier with a high slew rate push-pull amplifier (translinear G_m cell) such that $I_{SR} > I_{bias}$. A brief illustration of these methods are included in Fig. 3.19.

3.3   External Capacitor-Free LDO

Almost all LDOs described so far use an external capacitor C_{out} at the output of the LDO as shown in Fig. 3.6. With the development of advanced sub-micron CMOS technologies and the economic benefit of process scaling and integration, LDOs are now being used in large numbers on a system-on-chip (SoC) settings. For the conventional LDO topologies, each LDO would at least require an external capacitor and its dedicated output pin. The combined effect, therefore, includes a large number of external bulky components, more PCB board space, increased pin count of the IC, and higher Bill-of-Material (BoM). This goes against the trend in consumer portable electronics, which is pushing for higher integration and smaller form factor. It is also undesirable for future green electronic systems, which are characterized by their compactness and manufactureability.

In this section, the challenges and methods of designing an external capacitor-free LDO are discussed. A new approach that better addresses power, efficiency, supply voltage, and die area will be presented in the next section.

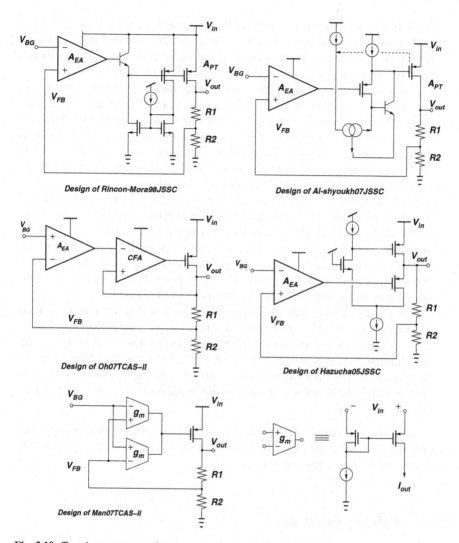

Fig. 3.19 Transient response enhancement techniques for a PMOS LDO

3.3.1   Challenges

The challenges for designing external capacitor-free LDO is best understood by studying the role of C_{out} in a typical LDO topology. There are at least three purposes for employing C_{out} in an LDO:

1. For many LDO implementations, C_{out} creates the external dominant pole $p_{out} = \frac{1}{R_{out}C_{out}}$ to stabilize the feedback loop, as described in Sect. 3.2.2. Although the pole position would still vary from heavy load to light load (due to changing

R_{out}), but R_{ESR} values are usually well chosen such that the ESR zero (or a tracking-zero, internal zero, etc.) will cancel out the amplifier dominant pole appropriately, or certain buffers were designed such that error amplifier no longer has a low frequency pole within the LDO Unity Gain Frequency (UGF).

2. C_{out} acts as a reservoir that temporarily sources and sinks current during load transients. As described in Sect. 3.2.3, it takes the LDO feedback loop at least Δt_1 to respond to any I_{load} change, during which the power MOSFET cannot change its output current, and hence the output voltage V_{out} would experience large overshot or undershot. During these instances, the capacitor C_{out} could charge or discharge itself to absorb or provide the excess transient current, as $I = C_{out} \frac{dV_{out}}{dt}$, before the LDO feedback loop responds.

   Notice that the contribution of C_{out} in reducing the transient undershot and overshot was not included in the analysis using Thevenin equivalent in Fig. 3.18. If included, when calculating the Thevenin equivalent circuit at AC, R_S will include the $1/sC_{out}$ impedance shunt to ground. For higher frequency sinusoid stimuli, which are exactly what an abrupt load transient generates, $R_S \to 0$ faster at the frequency of interest if a larger C_{out} is included. This is in agreement with the analysis above using only the characteristics of the capacitor.

3. C_{out} improves LDO PSRR. As explained in Thiele and Bayer (2005), at high frequencies, e.g. 100 kHz and beyond, the output impedance of the pass element and C_{out} forms a low pass filter that filters out high frequency supply ripple. The only limitation of this benefit is the ESR of C_{out}: the low pass filtering effect diminishes at very high frequency, e.g. 10 MHz and beyond when the C_{out} is shorted and the ESR and pass element output impedance forms a resistive divider (Gupta et al. 2004).

As a result, capacitor-free LDO design will face significant challenges in but not limited to stability (frequency compensation), load transient response, and PSR performance due to the absence of C_{out}.

3.3.2 Existing Solutions

Existing designs of an external capacitor-free LDO fall into two categories: two-stage approach and three-stage approach. In both approaches, the output power MOSFET is considered the last gain stage. Depending on the error amplifier structure (whether it is a single stage, or more precisely, single dominant pole structure, or a two-stage two-pole system), the division can be made accordingly.

Generally speaking, three-stage designs enjoy higher DC loop gain, therefore require less on-chip compensation capacitor (die area). However, multi-stage system are prone to introduce complex poles, which means even though the compensation capacitance can be small, certain conditions have to be guarded carefully. As will be discussed later, the output power MOSFET stage needs to be kept above minimum power level. Two-stage designs, on the other hand, require larger compensation capacitance, but would not have as many inherent poles and additional phase shift. It can be very attractive when power consumption is a premium, but the on-chip die area overhead is not negligible.

Leung and Mok (2003) first proposed a capacitor-free LDO based on a three-stage approach: Damping Factor (ζ) Control (DFC) method used in low voltage, multi-stage amplifier driving large capacitive load (Leung et al. 2000). DFC was an improvement on Nested Miller Compensation (NMC), a well known low voltage high gain multi-stage amplifier compensation method (Eschauzier et al. 1992; Huijsing 2000). Instead of nesting C_{m1} and C_{m2}, Leung et al. (2000) disconnected C_{m2} from the output node to reduce loading, which improved BW, and kept C_{m2} connected to the output of the first stage with an extra gain stage to boost C_{m2} effect, which reduced the magnitude peaking of the loop gain around unity gain frequency (UGF) from the complex pole pair. Lau et al. (2007) improved the method of Leung and Mok (2003) by an alternative connection of C_{m2} without the extra gain stage.

Milliken et al. (2007) proposed a capacitor-free LDO from a different perspective. If a single-stage error amplifier is used, topology wise, the LDO is very similar to a classic two-stage operational amplifier. Miller compensation, which is the most common compensation method for a two-stage opamp, can be easily applied. The difference is that load current varies by orders of magnitude, which means the gain of the second stage also varies significantly. To guarantee the effectiveness of the pole-splitting, a relative large on-chip compensation capacitance is needed to take advantage of the Miller effect. Milliken et al. (2007) proposed a "pseudo-differentiator", which effectively amplifies the capacitative feedback current. It also embodied Ahuja compensation method (Ahuja 1983), which removed the RHP zero by blocking the capacitative feed forward path while reducing capacitative loading for the input stage.

Methods of implementing an external capacitor-free LDO other than three-stage or two-stage approaches do exist. For example, Man et al. (2008) proposed a single-transistor-control (STC) LDO, which replaced the EA with a flipped voltage follower. Though the small-signal output impedance is reduced, which will improve the load transient response, a significant draw back is the lack of sufficient DC line regulation range. In order to keep all transistors in the saturation region,

$$V_{in} < V_{out} + V_{gs} - V_{ov} \tag{3.34}$$

which greatly compromised the line regulation range. As a result, STC-LDO and its various forms will not be discussed in details in this book. A brief illustration of these methods are included in Fig. 3.20.

3.4   Current-Area Efficient Capacitor-Free LDO

This section proposes an input current-differencing technique in designing a capacitor-free low-dropout regulator to simultaneously achieve sleep-mode efficiency and silicon real estate saving (Hu et al. (2010b)). With no minimum output

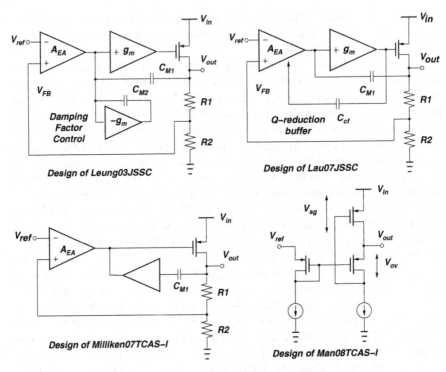

Fig. 3.20 Existing methods of designing external capacitor-free LDOs

current required to be stable, the regulator could greatly improve SoC efficiency during standby, which is extremely attractive for battery powered applications. Designed in TSMC 0.18 μm CMOS technology, it regulates 1.8 to 1.2 V supply down to 1 V with 100 mA maximum output current and can drive up to 100 pF of load parasitic capacitance. Compared with prior art with the same sleep-mode compatibility and similar output current range, it reduces on-chip compensation capacitance from 21 pF to 4.5 pF.

To further demonstrate the effectiveness of the proposed technique, two otherwise identical, 2.5-5 V to 2.3 V, 0-100 mA capacitor-free LDOs are designed and implemented in ON Semi 0.5 μm CMOS process available from MOSIS educational program, one using the conventional method, and the other using the proposed design (Hu and Ismail (2010)). Though both methods are effective when a minimum output current (stay alive current) is available, only the proposed LDO remains stable when stay alive is removed. This research enables CMOS capacitor-free LDOs to be truly robust and power efficient, ideal for future green electronic products.

3.4.1 Motivation: Efficiency Size Trade-off

As described in Sect. 3.3.2, a major limitation in prior arts on capacitor-free LDO is the trade-off between size and efficiency: In the two-stage approach, a significant amount of on-chip compensation capacitance (23 pF) was employed to compensate for full load range (0 - $I_{out,max}$) (Milliken et al. 2007). In the three-stage approaches, a minimum load current of $I_{out,min}$ ranging from 1 mA (Leung and Mok 2003) to 100 μA (Lau et al. 2007) is required for loop stability. This minimum current constantly drains the battery and degrades the system efficiency. Although 100 μA is small compared to $I_{out,max}$ of 100 mA, it is still larger than the SoC average sleep current in standby-power-critical applications, or the total quiescent current budget for a state-of-the-art battery management chip. Even if an LDO is not required to operate under extremely low I_{out}, in which cases the LDO itself will be shut down to save power (also known as hibernate mode), having a truly stable regulator at zero output current helps guard against unexpected load transient dips. Hence, a new capacitor-free topology that is both efficient and compact is needed.

3.4.2 Approach: Excessive Gain Reduction

The main observation from existing three-stage capacitor-free LDO designs is that not all gain contribute to the desired Nested Miller effect. DC gain in stages that do not contribute to Miller pole-splitting should be minimized, and if possible, be redistributed to stages that do. As a result, given a total DC gain budget to meet the LDO accuracy requirement, the stability of the LDO can be greatly improved (Hu and Ismail 2010).

The input differential amplifier, G_1, is an example of an excessive gain stage, as seen in Fig. 3.21a, which analyses the stability of a conventional three-stage nested-Miller capacitor-free LDO. G_1, G_2, and G_3 represents the input differential stage, high gain intermediate stage, and the common-source pass element output stage respectively. At least three parasitic poles are present in the loop: $p_1 = 1/(R_1 C_{M1} G_2 G_3)$, $p_2 = 1/(R_2 C_{M2} G_3)$, and $p_3 = g_{m3}/C_3$, where R_i, C_i and g_{mi} refer to the output resistance, parasitic capacitance, and transconductance of gain stage G_i ($i = 1, 2, 3$) respectively (Hu et al. 2010b).

Notice the gain G_1 does not contribute to creating the dominant pole p_1, or the pole-splitting of p_2 and p_3. Other than tracking the difference between the actual output V_{out} and the desired output V_{REF}, G_1 is an excessive gain stage that eats into the total DC gain $A_v(0)|_{dB} = G_1 + G_2 + G_3$ budget. Thus, an attenuating stage, $-G_1$ (in dB), can be introduced to reduce the excessive loop gain G_1. G_2, G_3 and g_{m3} can therefore be designed with larger values, enhancing the Miller effect by pushing p_1 to a lower value of p_1' and splitting p_2 and p_3 further apart to p_2' and p_3'.

Fig. 3.21 (a) Excessive gain in existing solutions and (b) Excessive gain reduction

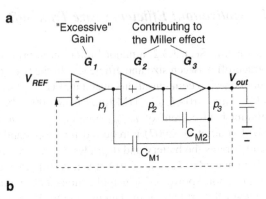

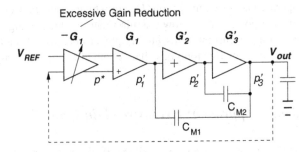

Ideally the attenuation should cancel the excess gain completely, leading to a unity gain differential amplifier. In practice, the attenuation might not be exactly equal to the excessive gain due to various forms of mismatch. However, a smaller than needed attenuation would still be helpful, as long as it does not add any additional parasitic pole, p^*, into the loop.

3.4.3   Method: Input Current-Differencing

The schematic of proposed capacitor-free LDO is shown in Fig. 3.22. Instead of using an operational amplifier as the error amplification block, it consists of current-differencing (CD) stage ($M_1 - M_6$), positive gain stage ($M_7 - M_{10}$), PMOS pass element (MPT), and resistive network (R_1, R_2). The CD stage is a CMOS version of the one proposed by Frederiksen et al. (1971), the original form of which is widely used in analog signal processing and frequency filtering (Keskin 2004). The LDO is internally compensated by MOSFET capacitance C_M and C_{cf} of 4.5 pF and 500 fF respectively. No external capacitor is needed. Other component values and transistor aspect ratios are listed in Fig. 3.22.

Input current-differencing technique was originally proposed in Frederiksen et al. (1971) for bipolar monolithic opamp design. As shown in Fig. 3.23a, a "current mirror" was added across the common-emitter input, resulting in a

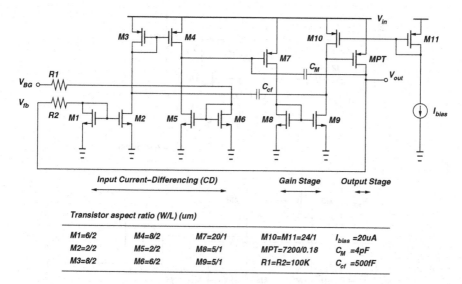

Transistor aspect ratio (W/L) (um)				
M1=6/2	M4=8/2	M7=20/1	M10=M11=24/1	I_{bias} =20uA
M2=2/2	M5=2/2	M8=5/1	MPT=7200/0.18	C_M =4pF
M3=8/2	M6=6/2	M9=5/1	R1=R2=100K	C_{cf} =500fF

Fig. 3.22 Schematic of the proposed capacitor-free LDO

current mode operation where the input currents are compared or differenced. With input resistance converting input voltage into current, a wide common-mode input range can be accommodated since both inputs are built-in biased with only $+V_{BE}$ above ground. A CMOS version, however, requires a third current mirror, as shown in Fig. 3.23, due to its zero gate current.

There are two major advantages when the technique is applied in an LDO. First of all, it allows low voltage operation. Notice that M_2, M_3, M_4 and M_6 can be viewed as an N-input differential pair, except for the removal of the tail current source due to the built-in biasing by the diode-connected M_1 and M_5. As a result, the minimum supply voltage could be as low as $V_{gs} + V_{ov} \approx 0.9 - 1.1 V$. In Lau et al. (2007), the conventional P-input differential pair would require $V_{gs} + 2V_{ov}$ or $V_{ref} + V_{gs} + V_{ov}$, whichever is higher, as shown in Fig. 3.23. Secondly, this configuration renders a smaller feedback factor β that helps reducing the excessive loop gain that threatens stability in zero load (Leung and Mok 2003; Lau et al. 2007). An alternative approach to lower loop gain would be reducing g_m, since changes in r_o shifts dc gain and the dominant pole simultaneously (constant GBW). Since $g_m = \frac{2I_D}{V_{ov}}$ in saturation region, reducing g_m would require either increasing V_{ov}, which is already limited in a low voltage situation, or lowering I_D, which is effective only to a certain extent before the input devices enter weak inversion where $g_m = \frac{2I_D}{V_{ov}}$ is no longer valid.

The stability of the proposed LDO can be analyzed as follows. Figure 3.24a shows that the LDO consists of a basic amplifier $a(j\omega)$ and a feedback network $f(j\omega)$ using two-port analysis. Figure 3.24b shows the gain magnitude versus frequency for the basic amplifier of the LDO, where different feedback factor f influencing the loop gain. The key to stabilizing the LDO is designing a small

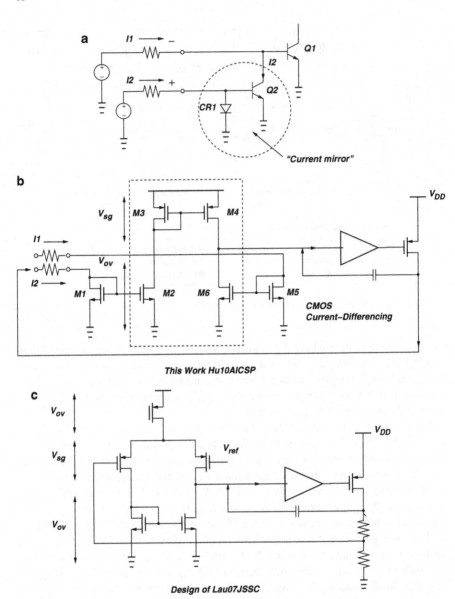

Fig. 3.23 Principle of input Current-Differencing: (**a**) bipolar implementation, (**b**) CMOS version, (**c**) advantages over prior arts

feedback factor $f = \frac{1/g_{mM1}}{1/g_{mM1}+R_2}$ in reducing the loop gain to avoid the residual $a(j\omega)$ peaking in addition to the Q-reduction compensation (Lau et al. 2007). The stability of the LDO can be analyzed in details with the small-signal equivalent circuit in Fig. 3.24c. Let g_{mMi} be the transconductance of transistor M_i ($i = 1, 2, \cdots, 10$). Let

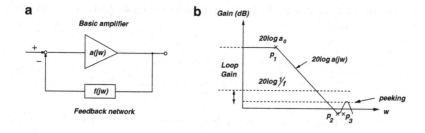

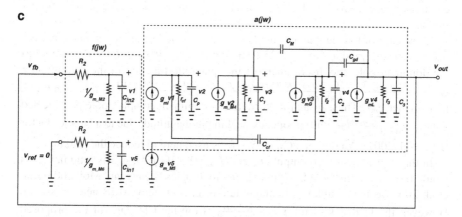

Fig. 3.24 LDO stability analysis. (**a**) Two-port model, (**b**) Bode plot, (**c**) small-signal equivalent circuit

g_{mI}, g_{mG} and g_{mL} be the transconductance for the CD stage ($M_1 - M_6$), gain stage ($M_7 - M_{10}$), and output stage (MPT) respectively. Let r_i, C_i ($i = 1, 2, 3$) be the output resistance and capacitance for each of the three stages. Let r_{cf}, C_p be the output resistance and capacitance associated with the drain of M_2, and C_{gd} be the gate-drain capacitance of MPT.

Assuming ideal frequency response of the Q-reduction current buffer ($M_3 - M_4$) (Lau et al. 2007), the loop gain of the LDO can be found by breaking the voltage feedback before R_2 and calculating the open-loop transfer function from V_{fb} to V_{out}:

$$L(s) = \frac{v_{out}(s)}{v_{fb}(s)}$$

$$= \frac{\frac{1/g_{mM1}}{1/g_{mM1}+R_2} \cdot g_{mI}g_{mG}g_{mL}r_1r_2r_3 \left\{ 1 + s\left(r_{cf}C_{cf} - \frac{C_{gd}}{g_{mL}}\right) - s^2\left(\frac{C_m(C_{gd}+C_2)}{g_{mG}g_{mL}} + \frac{C_{gd}r_{cf}C_{cf}}{g_{mL}}\right)\right\}}{(1 + sC_Mg_{mG}g_{mL}r_1r_2r_3)\left(1 + s\frac{C_mC_{gd}(g_{mL}-g_{mG})+C_{cf}C_3g_{mG}+C_mC_{cf}g_{mG}g_{mL}r_{cf}}{C_mg_{mG}g_{mL}} + s^2\frac{(C_{gd}+C_2+C_{cf})\cdot C_3}{g_{mG}g_{mL}}\right)}$$

$$(3.35)$$

Since the transconductance of the pass element g_{mL} varies greatly with the output current, the stability of the LDO should be analyzed in three different load conditions. In the case of both moderate and high output current ($I_{out} > 1mA$), the loop gain can be simplified as:

$$L(s) = \frac{v_{out}(s)}{v_{fb}(s)} = \frac{\frac{1/g_{mM1}}{1/g_{mM1}+R_2} \cdot g_{mI}g_{mG}g_{mL}r_1r_2r_3(1+s(r_{cf}C_{cf}))}{(1+s/p_1)\left(1+s\frac{C_{gd}+C_{cf}}{g_{mG}} + s^2\frac{(C_{gd}+C_2+C_{cf})\cdot C_3}{g_{mG}g_{mL}}\right)} \tag{3.36}$$

where $p_1 = 1/(sC_M g_{mG} g_{mL} r_1 r_2 r_3)$ is the dominant pole. Two real roots p_2, p_3 exist for the quadratic function as $(4(C_{gd}+C_2+C_{cf})\cdot C_3)/g_{mL} \le (C_{gd}+C_{cf})^2/g_{mG}$ holds due to a big g_{mL}. Zero $z_1 = 1/r_{cf}C_{cf}$ cancels p_2, and p_3 is located at a much higher frequency.

In the case of low output current ($I_{out} : 100\mu A - 1mA$), the loop gain expression needs to be analyzed in full as (3.35). Though $p_1 = 1/(sC_M g_{mG} g_{mL} r_1 r_2 r_3)$ remains the dominant pole, the discriminant of the quadratic function in the denominator $\Delta = [\frac{C_m C_{gd}(g_{mL}-g_{mG})+C_{cf}C_3 g_{mG}+C_m C_{cf}g_{mG}g_{mL}r_{cf}}{C_m g_{mG}g_{mL}}]^2 - \frac{4(C_{gd}+C_2+C_{cf})\cdot C_3}{g_{mG}g_{mL}}$ is less than 0 due to a smaller g_{mL}. Thus, p_2 and p_3 form a non-dominant complex pole pair located at $\omega_{2,3} = \sqrt{\frac{g_{mG}g_{mL}}{(C_{gd}+C_2)\cdot C_3}}$ with frequency peaking in magnitude due to a large Q-factor (Lau et al. 2007). An optional C_{cf} of 500 fF is used here to reduce the peaking.

In the case of near zero output current ($I_{out} : 0 - 100\mu A$), g_{mL} is minimal. This would result in an unstable LDO in existing topologies (Leung and Mok 2003; Lau et al. 2007) as the complex pole magnitude peak rises above 0 dB near crossover. However, the feedback factor $f = \frac{1/g_{mM1}}{1/g_{mM1}+R_2}$ from the CD stage of the proposed LDO lowers DC loop gain and effectively suppresses the peak below 0 dB even with zero output current.

3.4.4 Simulation Results

The proposed LDO has been simulated in TSMC 0.18 μm 1.8V/3.3V RFIC 1P6M+ process with key technology data listed in Table 3.1. It regulates 1.2–1.8 V input supply to fixed 1 V output with 51 μA quiescent current and 100 mA maximum outputs current. No minimum output current is required, i.e., the LDO is stable at $I_{out} = 0$. Line regulation at output $I_{out} = 0$ and $I_{out} = 100 mA$ are 0.223%/V and 0.728%/V respectively. Load regulation is 10.7 ppm/mA at $V_{in} = 1.2V$. Figure 3.25 shows the transient response at various load parasitic and process corner conditions. Worst-case 0.5% error recovery time is less than 13 μs under full load transient ($0 - 100mA$) of 1 μs rise and fall time. Power supply rejection ratio (PSRR) is − 41 dB at 1 kHz, and output noise is 6.3 $\mu V/sqrtHz$ and 640 $nV/\sqrt{Hz}$ at 100 Hz and 10 kHz respectively.

To demonstrate the advantage of proposed work, a brief comparison with existing work is shown in Table 3.2. This work reduces $I_{out,min}$ to zero (Lau et al. 2007) and compensation cap from 21 pF (Milliken et al. 2007) to 4.5 pF, which contributes

Table 3.1 TSMC 0.18 μm CMOS process technology (2 V nominal devices)

	1.8 V NMOS	1.8 V PMOS
$V_{th,0}$ (mV)	475	449
t_{ox} (nm)	4.08	4.08
C_{ox} (fF/μm^2)	8.46	8.46
$k' = 1/2\mu C_{ox}$ ($\mu A/V^2$)	340	70
Breakdown Voltage	1.8	1.8

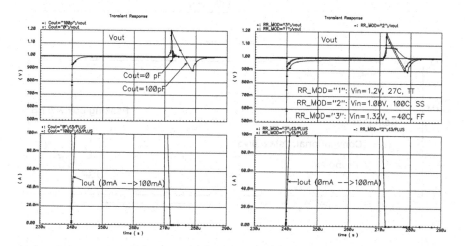

Fig. 3.25 Transient response of LDO under different output parasitic and corner conditions I_{out}: 20 mA/div., V_{out}: 100 mV/div., 10 us/div

Table 3.2 Performance comparison of the capacitor-free LDOs

	Lau et al. (2007)	Milliken et al. (2007)	This work, Hu et al. (2010b)
Process	CMOS 0.35	CMOS 0.35	CMOS 0.18
Year	2007	2007	2009
V_{in} (V)	1.2-3.3	3-4.2	1.2-1.8
V_{out} (V)	1	2.8	1
Settling time (μs)	Not reported	15	13
$I_{out,max}$ (mA)	100	50	100
$I_{out,min}$ (μA)	100	0	0
I_q (μA)	100	65	51
On-chip cap (pF)	6	21	4.5
Area (mm^2)	0.125	0.120	0.020
Area/L^2	1.020	0.980	0.617
FOM	0.120	0.273	0.027

greatly to the area saving. A figure of merit adapted from Hazucha et al. (2005) defined as $FOM = \frac{I_{out,min} + I_q}{I_{out,max}} \times \frac{C_{onchip}}{C_{out,par}}$ takes both effects into account. The smaller the FOM, the better current-area efficiency the LDO achieves.

Table 3.3 Capacitor-free
LDO design specifications

Category	Parameter	Specs
Voltage	V_{in}	2.5-5 V
	V_{out}	2.3 V
Current	$I_{out,min}$	0
	$I_{out,max}$	100 mA
Output	External	Not required
Capacitor	On-chip parasitic	0-100 pF
Efficiency	I_Q	$\leq 100\,\mu A$

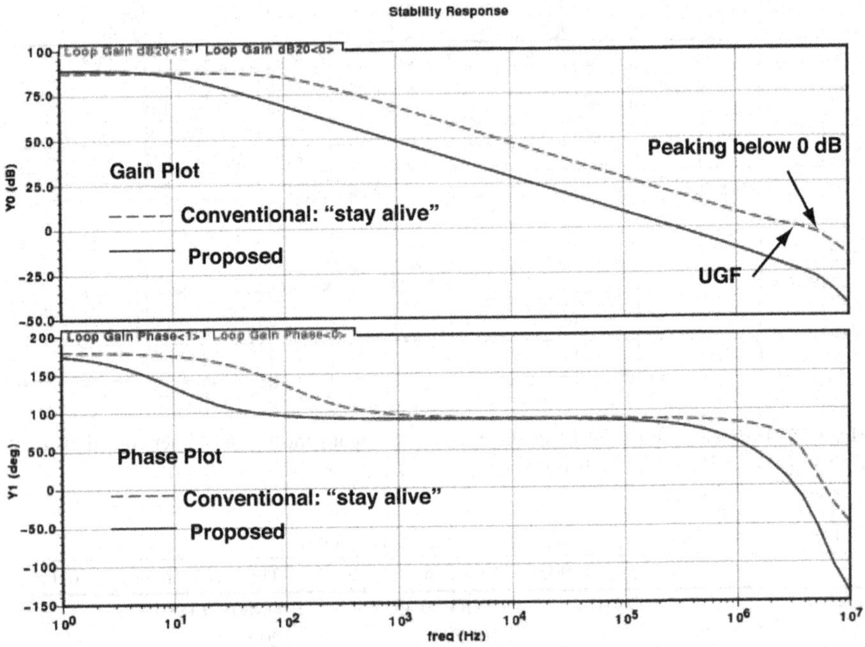

Fig. 3.26 Bode plots of the conventional and proposed LDOs at $I_{out,min} = 100\,\mu A$

Notice that in Table 3.2, the three LDOs are manufactured in different CMOS processes. Even though the FOM, as well as area comparison have all taken the process advancement into account, a stronger way to demonstrate the effectiveness of the proposed technique is to design two otherwise identical capacitor-free LDOs in the same process, one using existing techniques, and the other using the one proposed in Sect. 3.4.3.

The design specifications for the two LDOs are listed in Table 3.3. The available CMOS process from MOSIS educational service was the ON Semiconductor (previously AMIS) 0.5 μm C5N CMOS process (MOSIS 2011). The two LDOs are designed and placed side by side in 1.5 mm × 1.5 mm area within a DIP40 package.

Figure 3.26 shows the Bode plots of the conventional and proposed capacitor-free LDOs at $I_{out,min} = 100\,\mu A$. Notice that both LDO have the same DC gain

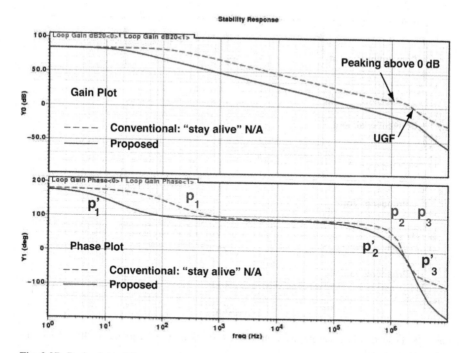

Fig. 3.27 Bode plots of the conventional and proposed LDOs at $I_{out,min} = 0$

of 84 dB, but the proposed LDO has a lower dominant pole and hence a smaller unity gain frequency (UGF) of 200 kHz instead of 2 MHz.[4] The phase margins are 60° (conventional) and 90° (proposed) respectively. Figure 3.27 shows the same Bode plots when the "stay alive" $I_{out,min}$ of 100 μA is not available. Notice that the conventional design is not stable because of the gain magnitude peaking, which leads to negative phase margin. The proposed design, on the other hand, is stable with a phase margin of 70°.

Figures 3.28 and 3.29 show the output voltage waveforms of the conventional and proposed LDOs in face of full-range load current transients (0–100 mA in 1 μs, Leung and Mok (2003), Lau et al. (2007), Milliken et al. (2007), Or and Leung (2010)). In Fig. 3.28, "stay alive" current $I_{out,min}$ of 100 μA is assumed, and both LDOs are stable, though they have various degrees of undershot and overshot. In Fig. 3.29, however, when "stay alive" current is not available, the conventional LDO oscillates, but the proposed LDO remains stable.

---

[4] A smaller UGF inevitably slows down loop transient response. However, the settling time of a LDO is usually limited by the slew rate at the gate of the pass element (Or and Leung 2010; Ho and Mok 2010). Thus, it is assumed here that the trade-off in UGF is acceptable, as long as the settling time does not deteriorate significantly (as seen later in Figs. 3.28 and 3.29). In addition, transient enhancement techniques reported in Or and Leung (2010), Ho and Mok (2010) can be always be applied as needed.

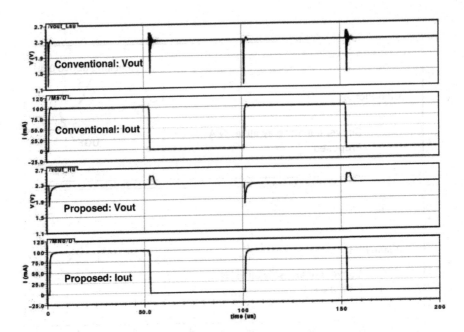

Fig. 3.28 Load transient waveforms with 100 μA of "stay alive" current. V: 400 mV/div, I: 25 mA/div, t: 50 μs/div

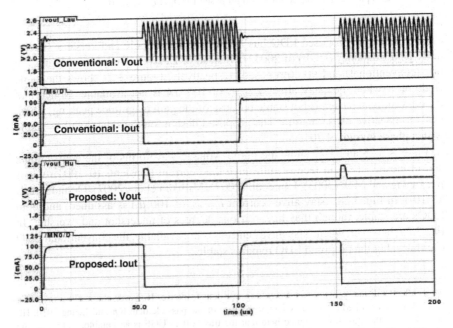

Fig. 3.29 Load transient waveforms without "stay alive" current. V: 200 mV/div, I: 25 mA/div, t: 50 μs/div

Table 3.4 Battery current saving from the proposed technique

	I_Q	$I_{out,min}$	$I_{batt,idle}$
Conventional	100 μA	100 μA	200 μA
Proposed	65 μA	0	65 μA
Saving			67.5%

The power efficiency of an LDO for portable application is often measured by its current efficiency during normal operation (Al-Shyoukh et al. 2007) and its total battery current drain during sleep (Hu et al. 2010b). Since the proposed technique would not require any "stay alive" current $I_{out,min}$ to combat zero-load oscillation, the total battery current drain during idle can be greatly reduced, as indicated in Table 3.4.

3.4.5   Measurement Results

This section lists the measurement results of the two identical LDOs designed and fabricated in the same package using MOSIS educational C5N process (Sect. 3.4.3). The bench test setup and PCB board design are briefly described. Due to a few top-level routing errors, some parts of the circuit fabricated were not able to function as intended. Studies and simulations were conducted to account for the simulation-measurement discrepancies.

3.4.5.1   Test Solutions Design

To effectively test the fabricated circuits, a complete set of test solutions, including on-chip test structure, chip-package interface, and PCB test board, need to be designed. In fact, IC testing is by no means trivial, and it has grown into a highly specialized field of its own (Burns and Roberts 2000). In industry, verification and production test represents up to 50–60% of the total time and cost in making Very Large Scale Integrated (VLSI) chips, making it the biggest single expense of the chip fabrication process (Demidenko et al. 2006). It was also predicted by the Semiconductor Industry Association in 2003 that by 2014, the cost of testing a transistor would exceed the cost of its manufacturing (Novak et al. 2007).

Unfortunately, the need for test engineering education has yet to be fully recognized in most schools and institutions at the time this book is written (Novak et al. 2007; Demidenko et al. 2006; Roberts 2008; Hudson and Copeland 2009; Hu et al. 2010a). Though certain approaches that do not require expensive automatic test equipment (ATE) have been proposed for mixed-signal IC testing (Hu et al. 2010a), the task of testing power management ICs with low cost and low complexity remains a challenge.

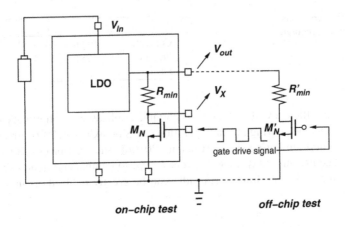

Fig. 3.30 On-chip test structure with off-chip option

The test solution design starts with the on-chip test structure. Since capacitor-free LDOs are most likely to be used to power on-chip loads, an on-chip load test structure (Or and Leung 2010) is included in the design, as seen in Fig. 3.30. A gate drive signal will turn on and off M_N periodically to introduce load current variation from 0 to $I_{out,max}$. V_{out} and V_X are directly monitored by the scope, while I_{out} is obtained indirectly as $\frac{V_{out}-V_X}{R_{min}}$. The drawback of this test structure is the variation of R_{min} after fabrication. As a result, an off-chip load test option with external high precision R'_{min} is also supported.

When the LDO cores are connected to the padding frames, Electrostatic Discharge (ESD) protection should also be considered. Without proper protection, human or machine touch of the packaged chip could introduce transient high voltage up to 1,000 V, which can easily damage the MOSFET gate or open-drain connected to the pad. In industrial products, sophisticated ESD protection circuits are designed for each type of input and ouput pad. In this design, however, two diode-connected MOSFETs were used for ESD protection. The schematic of the ESD is shown in Fig. 3.31. During normal operation, both the top PMOS and bottom NMOS are off, introducing no additional loading to the power rail. When the input voltage starts to rise above V_{DD}, the top PMOS would clamp the pad voltage to be one V_{SG} above V_{DD}. Likewise, the bottom NMOS would clamp the pad voltage to be one V_{GS} below V_{SS} (GND in this design). Both MOSFETs are designed to be sufficiently large to be able to carry the large transient current. The photo of the ESD structures after fabrication is shown in Fig. 3.32. Notice that there are two guard rings around the two power MOSFETs respectively. The N-type guard ring of the NMOS and the source of the PMOS should be connected to the highest voltage (V_{DD}), and the P-type guard ring for the PMOS and the source of the NMOS should be connected to the lowest voltage (V_{SS}).

As mentioned earlier, both the proposed and the existing LDOs are implemented on the same chip in ON Semi 0.5 μm CMOS process. A die photo of the "dual

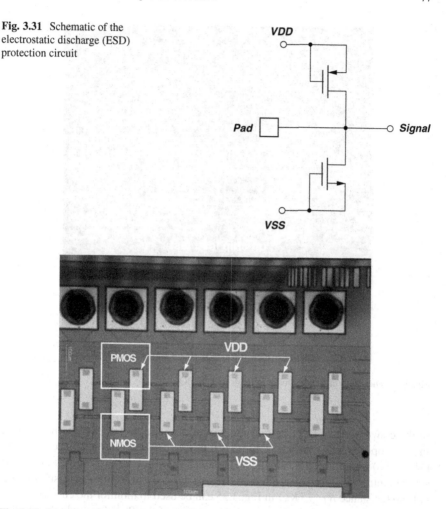

Fig. 3.31 Schematic of the electrostatic discharge (ESD) protection circuit

Fig. 3.32 Die photo showing the electrostatic discharge (ESD) structure

core" power management chip is shown in Fig. 3.33, which also shows the bond wires that connects the I/O pads to the package. The core of the two LDOs are laid side by side, each with their dedicated bias, power supply, and ground. A zoom-in view shows the on-chip resistance (implemented using High RES poly), capacitance (MOSFET capacitance), the pass element (multiple instance, multiple finger parallel connection), and the error amplifiers (Figure 3.34).

To test the packaged chip, a customized printed circuit board (PCB) was also designed, the schematic of which is shown in Fig. 3.35. Due to the lack of on-chip reference generation, a number of external stimuli other than V_{in} and GND are required, which are accommodated by test points. The external forced voltages are supposed to be tuned as the biasing conditions of key stages are monitored.

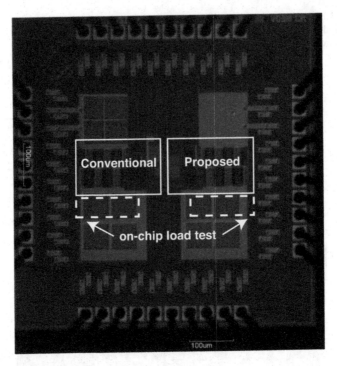

Fig. 3.33 Die photo of the dual-core capacitor-free LDO

In the conventional LDO, the biasing current of the input pair will be monitored. In the proposed LDO shown in Fig. 3.22, the biasing current of the second stage $(M_7 - M_{10})$ is monitored by having an identical MOSFET M_{iprb} as M_9 with gate and source connected. The open drain of M_{iprb} is pinned out to pin $I_{prb,2}$, such that a pull-up resistance $R_4 : 75\,K\Omega$ in Fig. 3.35 can be used to monitor the current.

Off-chip test structure takes up a large portion of the PCB area, which includes power NMOS ($Q_1 - Q_2$: ZVNL120A, $R_{ds,ON} = 10\,\Omega$, BV= 200 V, $0.5 < V_{th} < 1.5$, $P_{max} = 0.25\,W$) as switches and 10 Ω power resistors as loads. An external gate-drive signal can be applied to the gate of $Q_1 - Q_2$ to introduce the load variations. The presence and absence of the "stay alive" current was controlled by an additional load branch with $R_{5,8} = 23\,K\Omega$. At nominal output voltage (2.3 V), it would introduce a 100 μA "stay alive" current. The current can be removed by disconnecting jumper $J_{5,8}$. Power supply filtering capacitor was chosen as 0.1 μF. The photo of the PCB is shown in Fig. 3.36. The board size was 3′ by 5′.

Finally, certain equipment and meters are needed for testing, which include DC tunable power supplies, function generator, and oscilloscope for the least. A portable multimeter and a voltage meter are also helpful in the debugging process. The connection of major test equipment to the chip under test is shown in Fig. 3.37. Notice that the connection is specific for load transient response measurement,

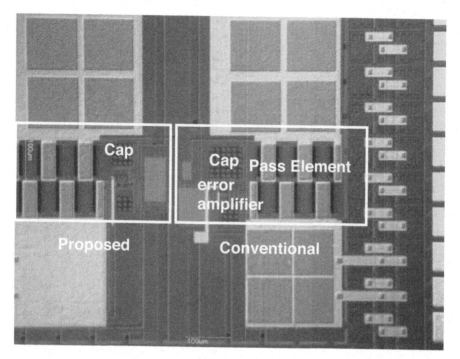

Fig. 3.34 Detailed die photo showing integrated circuit components: Resistors, Capacitors, Pass elements, and Error amplifiers

which is usually the key one which measures the LDO loop stability, speed, and transient response all-in-one. Slightly different connections (mostly at the load side) may be needed for other tests.

3.4.5.2 Measurements and Discrepancies

The measured load transient response of the conventional LDO (Lau et al. 2007) is shown in Fig. 3.38. The input voltage V_{in} is 3.7 V, and the nominal output is 3.469 V. When jumper J5 is connected, R5 provides the conventional LDO with approximately 150 μA of "stay alive" current. A periodic (100 kHz) square waveform ($V_{low} = 1 V, V_{high} = 2 V$) was applied at the gate of Q_1. According to the measurement at the test point U10 in Fig. 3.35 and the resistance value of R10, the load current change ΔI_{out} is as large as 137 mA. The output voltage dropped from 3.469 V to 3.094 V, which corresponds to a 375 mV drop. No oscillation or settling time is observed, which is in agreement with the analysis of this book and Lau et al. (2007), as well as the measurements of Lau et al. (2007).

The proposed LDO, however, suffers from over current ($I_{in} > 1A$) in a random amount of time after system powers up. The cause has been identified as missing

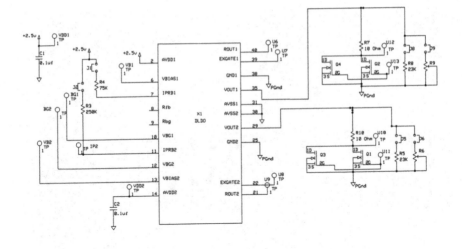

	Analog VLSI Lab	
	DLDO Test Board	
John Hu	Rev 1.0 8/9/2010	Page # or name

Fig. 3.35 Schematic of the printed circuit board (PCB)

vias at top-level interconnect. As seen in Fig. 3.39, three global interconnects that are supposed to pin out R_{fb}, $I_{prb,2}$, and $V_{BIAS,1}$ are in fact disconnected because of missing vias from Metal 1 to Metal 2. The disconnection of R_{fb} and $I_{prb,2}$ deprives the opportunity to monitor the attenuation factor $f = \frac{1/g_{mM1}}{1/g_{mM1}+R_2}$, and the biasing status of the second gain stage G_2 in Fig. 3.21. The disconnection of $V_{BIAS,1}$, however, is fatal, because the gate of M_{10} (in Fig. 3.22) is not defined, leaving the high impedance node of the drain of M_{10} and M_9 unpredictable, which could easily trigger a large amount of current flow in *MPT* by going below its V_{th}. The design error could have been avoided if padframe level layout-versus-schematic (LVS) CAD tool were available.

This analysis can be further verified by block level post-layout simulation with the disconnection of $V_{BIAS,1}$ properly modeled. A diagnostic test bench is shown in Fig. 3.40. The voltage at the reference pin $V_{BIAS,1}$ was set as a DC variable ranging from the supply to ground. A DC sweep on the $V_{BIAS,1}$ from 0 to 2.5 V under 3.5 V input is shown in Fig. 3.41, where the total battery current (negative in polarity) is monitored. If $V_{BIAS,1}$ were properly connected to the external bias, it would be very close to 2.5 V, approximately one V_{SG} below V_{in} to properly turn on the power MOSFET. The battery current would be equal to the load current in the diagnostic test bench in Fig. 3.40, which is on the order of 100 μA. However, as the simulation in Fig. 3.41 shows, the battery current increases dramatically as the voltage at $V_{BIAS,1}$

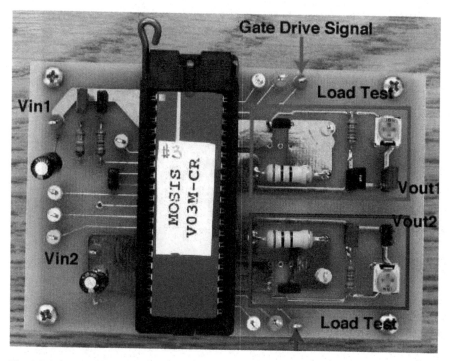

Fig. 3.36 Photo of the PCB

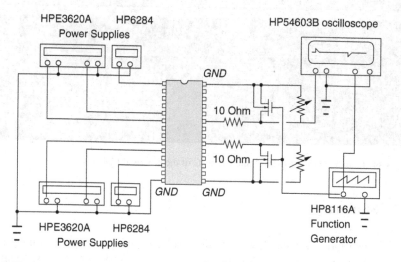

Fig. 3.37 Connection of the test equipment

is reduced, leading to potentially 5X increase in schematic level simulation, even more after post-layout extraction. This qualitatively explains the 50-50 chance of input current surge observed during silicon measurements.

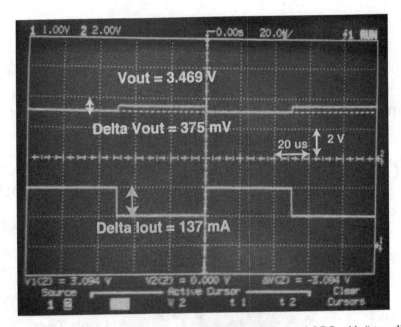

Fig. 3.38 Measured transient response of the fabricated conventional LDO with "stay alive" current

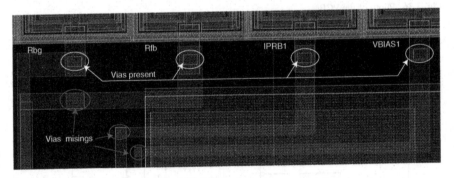

Fig. 3.39 Layout errors at top-level interconnects of the proposed LDO core

3.5 Summary

This chapter looks into the design of the most basic power management integrated circuit block, a linear regulator. Before exploring any design, a power supply perspective was established, which is that an optimum power management circuit should approximate an ideal voltage source.

With this perspective, all four feedback configurations were studied. Only the two output shunt feedbacks are feasible for a voltage source. Depending on whether the

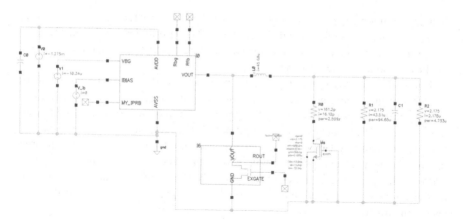

Fig. 3.40 Diagnostic test bench to analyze the measurement-simulation discrepancies

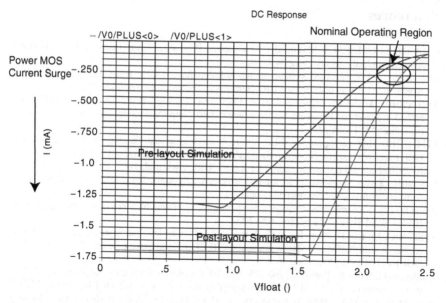

Fig. 3.41 Schematic and Post-layout simulations of the proposed LDO with disconnected $V_{BIAS,1}$ that correlate with measurement results

feedback signal is voltage or current, series-shunt and shunt-shunt linear regulators can be constructed. Though shunt-shunt regulators are rarely reported in literature, some examples of which are provided.

The chapter then went on to discuss series-shunt linear regulators in details. Common-drain high drop-out (HDO) and common-source low drop-out (LDO) configurations are identified as two major circuit topologies, and the pros and cons of each topology is explored. Due to the efficiency advantage, the LDO topology is more common for portable applications, and the key performance matrix of an LDO is then discussed in details, which includes various DC, AC, and transient performance parameters. A brief review of literature on existing methods to improve on those parameters are also presented.

Due to the growing trend of high-level integration and the strict requirement for small footprint in portable devices, external capacitor-free LDOs and their unique design challenges are also studied. A sleep-mode current-area efficient LDO using input current-differencing technique is proposed. The motivation stems from the fact that existing cap-free LDOs either sacrifice sleep-mode efficiency or require large on-chip die area. The proposed design reduces the excessive loop gain, which help addresses the current-area trade-off at sleep mode. The simulation and measurement results are also presented with analysis on the discrepancies between measurement and simulation data.

References

OMA (2008) OMAP3430 Multimedia Applications Processor. Texas Instruments, URL http://focus.ti.com/pdfs/wtbu/ti_omap3430.pdf

TIC (2009) A USB enabled system-on-chip solution for 2.4-GHz IEEE 802.15.4 and Zigbee Applications. Texas Instruments, URL http://focus.ti.com/lit/ds/symlink/cc2531.pdf

MC1 (2010) Advanced Zigbee-compliant Platform-in-Package (PiP) for the 2.4 GHz IEEE 802.15.4 Standard. Freescale Semiconductor, URL http://www.freescale.com/files/rf_if/doc/data_sheet/MC1322x.pdf

STM (2010) High-performance, IEEE 802.15.4 wireless system-on-chip. STMicroelectronics, URL http://www.st.com/stonline/products/literature/ds/16252/stm32w108cb.pdf

NRE (2010) National renewable energy laboratory. URL www.nrel.gov

TPS (2010) Power Management IC for Li-Ion Powered Systems. Texas Instruments, URL http://focus.ti.com/lit/ds/symlink/tps65023.pdf

Ahuja B (1983) An improved frequency compensation technique for cmos operational amplifiers. IEEE J Solid-State Circuits 18(6):629–633

Al-Shyoukh M, Lee H, Perez R (2007) A transient-enhanced low-quiescent current low-dropout regulator with buffer impedance attenuation. IEEE J Solid-State Circuits 42(8):1732–1742, DOI 10.1109/JSSC.2007.900281

den Besten G, Nauta B (1998) Embedded 5 V-to-3.3 V voltage regulator for supplying digital IC's in 3.3 v cmos technology. IEEE J Solid-State Circuits 33(7):956–962, DOI 10.1109/4.701230

Burns M, Roberts GW (2000) An Introduction to Mixed-Signal IC Test and Measurement. Oxford University Press, USA

Camacho D, Gui P, Moreira P (2009) An NMOS low dropout voltage regulator with switched floating capacitor gate overdrive. In: 2009 MWSCAS. 52nd IEEE Int. Midwest Symp. Circuits and Systems, pp 808–811, DOI 10.1109/MWSCAS.2009.5235891

Chava C, Silva-Martinez J (2004) A frequency compensation scheme for ldo voltage regulators. IEEE Trans Circuits Syst I, Reg Papers 51(6):1041–1050, DOI 10.1109/TCSI.2004.829239

Demidenko S, Lai V, Kassim Z (2006) Industry-academia collaboration in undergraduate test engineering unit development. In: Proc. 3rd IEEE Int. Workshop Electron. Design Test and Applicat. (DELTA'06), pp 116–122, DOI 10.1109/DELTA.2006.58

Dokania R, Rincon-Mora G (2002) Cancellation of load regulation in low drop-out regulators. Electron Letters 38(22):1300–1302

El-Nozahi M, Amer A, Torres J, Entesari K, Sanchez-Sinencio E (2010) High psr low drop-out regulator with feed-forward ripple cancellation technique. IEEE J Solid-State Circuits 45(3):565 –577, DOI 10.1109/JSSC.2009.2039685

Eschauzier R, Kerklaan L, Huijsing J (1992) A 100-mhz 100-db operational amplifier with multipath nested miller compensation structure. IEEE J Solid-State Circuits 27(12):1709–1717, DOI 10.1109/4.173096

Frederiksen T, Davis W, Zobel D (1971) A new current-differencing single-supply operational amplifier. IEEE J Solid-State Circuits 6(6):340–347

Giustolisi G, Falconi C, D'Amico A, Palumbo G (2009) On-chip low drop-out voltage regulator with NMOS power transistor and dynamic biasing technique. Analog Integr Circuits Signal Process 58(2):81–90, DOI http://dx.doi.org/10.1007/s10470-008-9234-1

Gray PR, Hurst PJ, Lewis SH, Meyer RG (2001) Analysis and Design of Analog Integrated Circuits, 4th edn. John Wiley & Sons, Inc.

Gupta V, Rincon-Mora G (2007) A 5ma 0.6m cmos miller-compensated ldo regulator with -27db worst-case power-supply rejection using 60pf of on-chip capacitance. In: IEEE ISSCC Dig. Tech. Papers, pp 520–521, DOI 10.1109/ISSCC.2007.373523

Gupta V, Rincon-Mora G, Raha P (2004) Analysis and design of monolithic, high psr, linear regulators for soc applications. In: Proc. IEEE 2004 Int. SoC Conf., pp 311–315, DOI 10.1109/SOCC.2004.1362447

Hazucha P, Karnik T, Bloechel B, Parsons C, Finan D, Borkar S (2005) Area-efficient linear regulator with ultra-fast load regulation. IEEE J Solid-State Circuits 40(4):933–940, DOI 10.1109/JSSC.2004.842831

Ho E, Mok P (2010) A capacitor-less cmos active feedback low-dropout regulator with slew-rate enhancement for portable on-chip application. IEEE Trans Circuits Syst II, Express Briefs 57(2):80 –84, DOI 10.1109/TCSII.2009.2038630

Hoon S, Chen S, Maloberti F, Chen J, Aravind B (2005) A low noise, high power supply rejection low dropout regulator for wireless system-on-chip applications. In: Proc. IEEE 2005 Custom Integrated Circuits Conf., pp 759–762, DOI 10.1109/CICC.2005.1568779

Hu J, Ismail M (2010) A true zero-load stable cmos capacitor-free low-dropout regulator with excessive gain reduction. In: Proc. Int. Conf. Electron. Circuits Syst. (ICECS), pp 978–981, DOI 10.1109/ICECS.2010.5724677

Hu J, Haffner M, Yoder S, Scott M, Reehal G, Ismail M (2010a) Industry-oriented laboratory development for mixed-signal IC test education. IEEE Trans Educ 53(4):662–671

Hu J, Liu W, Ismail M (2010b) Sleep-mode ready, area efficient capacitor-free low-dropout regulator with input current-differencing. Analog Integr Circ Sig Process 63(1):107–112

Hudson T, Copeland B (2009) Working with industry to create a test and product engineering course. In: IEEE Int. Conf. Microelectron. Syst. Educ. (MSE'09), pp 130–133

Huijsing JH (2000) Operational Amplifiers: Theory and Design, 1st edn. Springer

Ingino J, von Kaenel V (2001) A 4-ghz clock system for a high-performance system-on-a-chip design. IEEE J Solid-State Circuits 36(11):1693–1698, DOI 10.1109/4.962289

Johns DA, Martin K (1997) Analog Integrated Circuit Design, 1st edn. John Wiley & Sons, Inc.

Keskin AU (2004) A four quadrant analog multiplier employing single CDBA. Analog Integr Circ Sig Process 40(1):99–101

King BM (2000) Advantages of using pmos-type low-dropout linear regulators in battery applications. Tech. rep., Texas Instruments, URL http://focus.ti.com.cn/cn/lit/an/slyt161/slyt161.pdf

Kruiskamp W, Beumer R (2008) Low drop-out voltage regulator with full on-chip capacitance for slot-based operation. In: 2008 ESSCIRC. 34th European Solid-State Circuits Conf., pp 346–349, DOI 10.1109/ESSCIRC.2008.4681863

Kwok K, Mok P (2002) Pole-zero tracking frequency compensation for low dropout regulator. In: Circuits and Systems, 2002 IEEE International Symposium on, IEEE, vol 4, pp IV–735

Lau SK, Mok PKT, Leung KN (2007) A low-dropout regulator for SoC with Q-reduction. IEEE J Solid-State Circuits 42(3):658–664, DOI 10.1109/JSSC.2006.891496

Lee BS (1999a) Technical review of low dropout voltage regulator operation and performance. Tech. rep., Texas Instruments, URL http://focus.ti.com/lit/an/slva072/slva072.pdf

Lee BS (1999b) Understanding the terms and definitions of ldo voltage regulators. Tech. rep., Texas Instruments, URL http://focus.ti.com/lit/an/slva079/slva079.pdf

Leung KN, Mok P (2003) A capacitor-free cmos low-dropout regulator with damping-factor-control frequency compensation. IEEE J Solid-State Circuits 38(10):1691–1702, DOI 10.1109/JSSC.2003.817256

Leung KN, Mok P, Ki WH, Sin J (2000) Three-stage large capacitive load amplifier with damping-factor-control frequency compensation. IEEE J Solid-State Circuits 35(2):221–230, DOI 10.1109/4.823447

Lin HC, Wu HH, Chang TY (2008) An active-frequency compensation scheme for cmos low-dropout regulators with transient-response improvement. IEEE Trans Circuits Syst II, Express Briefs 55(9):853–857, DOI 10.1109/TCSII.2008.924366

Man TY, Mok P, Chan M (2007) A high slew-rate pushpull output amplifier for low-quiescent current low-dropout regulators with transient-response improvement. IEEE Trans Circuits Syst II, Express Briefs 54(9):755–759, DOI 10.1109/TCSII.2007.900347

Man TY, Leung KN, Leung CY, Mok P, Chan M (2008) Development of single-transistor-control ldo based on flipped voltage follower for soc. IEEE Trans Circuits Syst I, Reg Papers 55(5):1392–1401, DOI 10.1109/TCSI.2008.916568

Milliken R, Silva-Martinez J, Sanchez-Sinencio E (2007) Full on-chip cmos low-dropout voltage regulator. IEEE Trans Circuits Syst I, Reg Papers 54(9):1879–1890, DOI 10.1109/TCSI.2007.902615

Mohan N, Undeland TM, Robbins WP (2003) Power Electronics: Converters, Applications and Design, 3rd edn. John Wiley & Sons, Inc.

MOSIS (2011) On semiconductor c5 process. URL http://www.mosis.com/on_semi/c5/

Novak F, Biasizzo A, Bertrand Y, Flottes ML, Balado L, Figueras J, Di Carlo S, Prinetto P, Pricoli N, Wunderlich HJ, Van Der Hayden JP (2007) Academic network for microelectronic test education. Int J Eng Educ 23(6):1245–1253

Oh W, Bakkaloglu B (2007) A CMOS low-dropout regulator with current-mode feedback buffer amplifier. IEEE Trans Circuits Syst II, Express Briefs 54(10):922–926, DOI 10.1109/TCSII.2007.901621

Oh W, Bakkaloglu B, Wang C, Hoon S (2008) A CMOS low noise, chopper stabilized low-dropout regulator with current-mode feedback error amplifier. IEEE Trans Circuits Syst I, Reg Papers 55(10):3006–3015, DOI 10.1109/TCSI.2008.923278

Or PY, Leung KN (2010) An output-capacitorless low-dropout regulator with direct voltage-spike detection. IEEE J Solid-State Circuits 45(2):458–466, DOI 10.1109/JSSC.2009.2034805

Razavi B (2001) Design of Analog CMOS Integrated Circuits, 1st edn. McGraw-Hill

Rincon-Mora G, Allen P (1998a) A low-voltage, low quiescent current, low drop-out regulator. IEEE J Solid-State Circuits 33(1):36–44, DOI 10.1109/4.654935

Rincon-Mora G, Allen P (1998b) Optimized frequency-shaping circuit topologies for ldos. IEEE Trans Circuits Syst II, Analog and Digital Signal Processing 45(6):703–708, DOI 10.1109/82.686689

Rincon-Mora GA (1996) Current efficient, low voltage, low-dropout regulators. PhD thesis, Georgia Institute of Technology

Roberts G (2008) ECSE-435B: Introduction to Mixed-Signal Test Techniques. McGill University, URL http://www.ece.mcgill.ca/~grober4/ROBERTS/COURSES/EE435/ee435.html

Salerno DC, Jordan MG (2006) Methods and circuits for programmable automatic burst mode control using average output current

Shi C, Walker B, Zeisel E, Hu B, McAllister G (2007) A highly integrated power management ic for advanced mobile applications. IEEE J Solid-State Circuits 42(8):1723–1731, DOI 10.1109/JSSC.2007.900284

Steyaert M, Sansen W (1990) Power supply rejection ratio in operational transconductance amplifiers. IEEE Trans Circuits Syst 37(9):1077–1084, DOI 10.1109/31.57596

Teel JC (2005) Understanding noise in linear regulators. Tech. rep., Texas Instruments, URL http://focus.ti.com/lit/an/slyt201/slyt201.pdf

Thiele G, Bayer E (2005) Current-mode ldo with active dropout optimization. In: 2005 IEEE Power Electronics Specialists Conf., pp 1203–1208, DOI 10.1109/PESC.2005.1581782

Wang X, Bakkaloglu B (2008) Systematic design of supply regulated LC-tank voltage-controlled oscillators. IEEE Trans Circuits Syst I, Reg Papers 55(7):1834–1844, DOI 10.1109/TCSI.2008.918004

Wong K, Evans D (2006) A 150ma low noise, high psrr low-dropout linear regulator in 0.13m technology for rf soc applications. In: Proc. ESSCIRC 2006 European Solid-State Circuits Conf., pp 532–535, DOI 10.1109/ESSCIR.2006.307507

Chapter 4
Switching Converters

With all the advantages in their simplicity and quality, linear regulators suffer from a significant drawback: efficiency. The efficiency of a linear regulator is upper bounded by its output-input voltage ratio, V_{out}/V_{in} (see Chap. 3). For example, if only half of the input voltage is needed at the output, the efficiency of a linear power supply would never exceed 50%, regardless of how the circuit is designed. This inefficiency is due to the usage of voltage division. The pass element and the load forms linear voltage divider (as seen in Fig. 3.5), in which the pass element always carries the total load current with $V_{DO} = V_{in} - V_{out}$ across. Hence, as much as $(V_{in} - V_{out}) \times I_{out}$ is dissipated. This limitation in efficiency and the power loss in the pass element becomes increasingly unacceptable for highly efficient portable and green electronic products.

Switch-mode DC-DC converters, on the other hand, do not rely on voltage division to generate the desired output. Capacitors (C) and inductors (L) are inserted into the power path as energy storage elements. Regular switching of semiconductor switches periodically reconfigures the power path, such that power is first harvested onto energy storage elements before it is delivered to the output. Since neither L, C nor switches will dissipate any power[1], the efficiency can reach 100% in theory (Mohan et al. 2003).

In many high power applications, input-output isolation is often required. This can be achieved by using a transformer. However, these bulky transformers are not always necessary for low power consumer applications. With high efficiency and maximum compactness as primary goals, the design of integrated non-isolated switch-mode DC-DC converters (with output power less than a few Watts) is of particular interest in this book. Relatively moderate output power levels also help reduce the thermal design efforts in packaging and heat sinks.

---

[1]Ideal switches do not dissipate any power: when the switch is ON, high current (I) flows through the switch with zero voltage (V) across; when the switch is OFF, no current flows through the switch, regardless of the voltage across it. Thus, $I \times V$ product is zero at all times.

J. Hu and M. Ismail, *CMOS High Efficiency On-chip Power Management*, Analog Circuits and Signal Processing, DOI 10.1007/978-1-4419-9526-1_4, © Springer Science+Business Media, LLC 2011

One of the challenges in these low power settings, however, is the need for high efficiency across a wide output load range. As discussed in Chap. 2, multiple power modes are common in portable systems to prolong battery lives. Even though switch-mode DC-DC converters are often preferred over LDOs because they are more efficient, this advantage can diminish at light load, when systems enter sleep modes.

In this chapter, the challenges in designing light-load efficient DC-DC converters are first discussed. Then, existing methods of light-load efficiency boosting are reviewed. Finally, a Long-Sleep Model (LSM) that improves on existing methods and facilitates design re-use is proposed. Simulation data are also presented.

4.1   Light-Load Efficiency Challenge

Ideally, a switch-mode DC-DC converter is lossless and the efficiency is 100% regardless of operations. In practice, ideal switches, inductors, and capacitors can only be approximated, an the control of the converter circuits requires static power consumption that can only be minimized. Therefore, the overall converter efficiency is degraded. Furthermore, the efficiency of any particular DC-DC converter is not constant. It varies with the mode of operation, output power, input voltage, and other factors.

Empirically, the efficiency of a DC-DC converter is a function of the output power[2] (Zhou and Rincon-Mora 2006). For any fixed mode of operation, the efficiency of a buck converter peaks at moderate output power and declines at both high and low output power, as seen in Fig. 4.1.

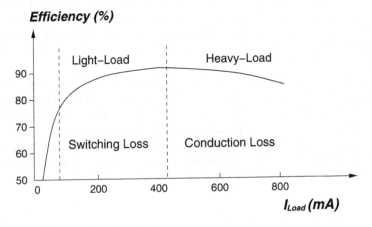

Fig. 4.1 Efficiency as a function of the output power (load current)

---

[2]In the case of fixed output voltage, the efficiency is a function of the load current.

A qualitative analysis can be conducted as follows. The power consumption of a DC-DC converter can be analyzed as:

$$P_{in} \approx P_{out} + P_{sw} + P_{cond} + P_{sht} + P_Q \qquad (4.1)$$

$P_{sw}, P_{cond}, P_{sht}$ and P_Q represent the switching loss, conduction loss, shoot-through power loss, and the circuit static power loss respectively (Ma et al. 2004). As the name implies, the conduction loss P_{cond} is due to the non-zero resistance of power transistors as well as other equivalent resistance in the power path. It is proportional to I_{out}. As I_{out} decreases, P_{cond} is reduced accordingly. Strictly speaking, the conduction loss is tied to the rooted mean-square current, i.e. $P_{cond} \propto I_{out,rms}$. As a result, the average current $I_{out,avg}$ is not the only factor in determining P_{cond}. Ripple current in the inductor can also impact P_{cond}, contributing to the power loss even when $I_{out,avg} = 0$.

The shoot-through power loss P_{sht} is incurred when a synchronous rectifier topology is used and the high-side and low-side power transistors are ON simultaneously. P_{sht} is relatively independent of I_{out}, and it can be reduced by proper dead-time control (also known as break-before-make) circuits (Huang et al. 2007). As a result, this power loss factor may or may not manifest itself on the right hand side of (4.1) depending on the operation of the converter.

The switching loss P_{sw} includes all power loss associated with switching on and off the power transistors. A big chuck of the switching loss is related to charging and discharging the huge power MOSFET, also known as the driving loss P_{dr}. This loss is proportional to gate capacitance (a combination of C_{gs} and C_{gb}) of the MOSFET, the switching frequency f_s, and the gate driving voltage swing, i.e. $P_{dr} \propto C_{gate} V_{string} \cdot f_s$. For simple rail-to-rail gate drive, $V_{swing} = V_{DD}$.

For non-resonant DC-DC converters, P_{sw} also includes hard switching loss. Hard switching loss refers to the loss incurred when the voltage across a switching element is non-zero, but the current through the element was abruptly switched from a non-zero value to zero, or vice versa. Regardless of how short this period is, there will be a non-zero voltage-current overlap, leading to power dissipation in the switch.

Conventional switch-mode DC-DC converters, such as Buck, Boost, Buck-boost, and Cuk topologies are all hard-switching converters, in other words there will be non-zero voltage across a switch when it is switched on or off. Thus, the voltage-current overlap is inevitable. Resonant converters, on the other hand, make use of RLC resonance circuit to implement zero-voltage switching. Compared with hard-switching DC-DC converters, this would require an additional capacitative element to form the RLC resonance tank (Mohan et al. 2003). This extra external components sometimes can not be justified in portable applications, where size is a premium. Thus, this book will assume that one of the canonical DC-DC topologies is used.

Last but not the least, P_{sw} includes the reverse recovery loss P_{rr} of diodes, whether it is a free-wheeling diode or the body diode of the low-side synchronous MOSFET. For each switching cycle, as much as Q_{rr} amount of energy is dissipated when a reversed-biased diode is switched to forward conducting state, hence the name

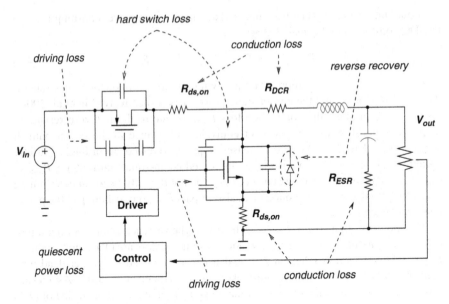

Fig. 4.2 Summary of power loss in a DC-DC converter

reverse recovery. In fact, $P_{rr} = V_{in}Q_{rr} \cdot f_s$. To reduce Q_{rr}, a Schottky diode can be used instead of a silicon diode. Schottky diode is a metal-semiconductor junction which has lower forward voltage drop and faster reverse recovery.

Therefore, all three components of P_{sw} show dependency on the switching frequency f_s. Depending on the operating method of the converter, the switching frequency f_s can be constant or variable. If a fixed f_s is used, then P_{sw} is independent of I_{out}. If an f_s proportional to I_{out} is used, P_{sw} scales down with I_{out} in general.

The static power loss P_Q includes all the quiescent power consumption in a DC-DC converter, such as the quiescent power consumed in various voltage and current references, amplifiers, and oscillators. Unless some dynamic biasing strategies are used, P_Q does not usually scale with I_{out}. In fact, P_Q remains constant in many DC-DC converters. P_Q rarely plays a dominant role in determining the power converter's efficiency at moderate to heavy load, as it is often minimal compared to other power losses, but it can become a limiting factor when I_{out} is reduced dramatically. A summary of the different power losses discussed so far and their associated circuit components are illustrated in Fig. 4.2 (Qahouq et al. 2007).

A similar analysis can be done to study the heavy load efficiency decline, in which case P_{cond} significantly increases and often becomes the dominant form of power loss. The heavy load efficiency decline is of great concern in high power applications, such as residential and utility power converters. It generates excessive heat that can over stress the components, shorten the life span of power equipment, and increase the cost of power delivery, which all eat into energy provider's bottom line.

This chapter primarily focuses on the light-load efficiency challenge that is more applicable to portable and battery-powered applications. Readers interested in high power converter designs may refer to Mohan et al. (2003) and Erickson and Maksimović (2001).

4.2 Existing Light-Load Efficiency Boosting Techniques

This section will briefly review some existing light-load efficiency boosting techniques. From the analysis in Sect. 4.1, each term on the right hand side of (4.1) demonstrates a different degree of dependence on the output power I_{out}. Among them, P_{sw} and P_Q can be a weak function of I_{out}. As I_{out} decreases, P_{sw} and P_Q do not scale back proportionally, if at all. Therefore, there are two main approaches to boost light-load efficiency: reducing P_{sw} and reducing P_Q as I_{out} diminishes.

From the analysis in Sect. 4.1, all three components of the switching loss P_{sw} are proportional to the switching frequency f_s. If we write P_{sw} as

$$P_{sw} = f_s \times E_{sw} \tag{4.2}$$

in which E_{sw} is the total switching-related energy loss in a single switching period, two sub-approaches can be identified: reducing the switching frequency f_s, or reducing the one-time switching loss E_{sw}.

Methods of reducing f_s include variable-frequency operation (Arbetter et al. 1995), where buck converter runs in discontinuous-conduction mode (DCM), hybrid mode (Zhou et al. 2000), or automatic Mode Hopping (MH) (Zhou and Rincon-Mora 2006). In these methods, the converter operates in synchronous continuous-conduction mode (CCM) during heavy load, but switches to various other operation modes as load decreases, such as asynchronous discontinuous-conduction mode (DCM).

One difficulty in the practical implementation of these methods has to do with automatic mode switching. Arbetter et al. (1995) relied on manual mode switch while Zhou and Rincon-Mora (2006) used empirical data to pre-set the boundaries of different modes. Zhou et al. (2000) realized automatic mode switch by monitoring $R_{ds,on}$-induced duty cycle drift: a larger I_{out} would introduce more duty cycle D deviation from the ideal $D = \frac{V_o}{V_i}$, if $R_{ds,on}$ of both the high-side and low-side power MOSFETs are non-negligible.

Another difficulty related to the use of variable f_s is the potential increase in inductor current ripples. If the switching frequency reduction is applied alone without operating mode adjustment or mode-hopping, the inductor current ripple could increase drastically. Larger inductor current ripple increases the RMS inductor current and its conduction loss. Therefore, it is possible that this increase in P_{cond} offsets the power saving from the reduced P_{sw}. Qahouq et al. (2007) studied the relationship between current ripple and frequency scaling methods and proposed a non-linear f_s scaling technique that preserves the steady-state voltage ripple to maintain good dynamic performance while keeping the efficiency benefits of variable f_s.

Methods of reducing E_{sw} also include two sub methods. As analyzed in Sect. 4.1, a large portion of E_{sw} is the gate driving loss E_{dr}, and it can be expressed as

$$E_{dr} = \frac{1}{2} C_{gate} V_{swing}^2 \qquad (4.3)$$

Therefore, either reducing C_{gate} or V_{swing} will reduce E_{dr} effectively. Power FET segmentation , or optimum MOSFET width control, is a method to reduce C_{gate}. In this method, the power transistor is implemented with segments of small transistors connected in parallel. All the segments are activated only at peak load power to achieve the minimum conduction loss. For a lighter load, only a few segments of transistors are activated, while the others are turned off to reduce the driving loss (Ma et al. 2004; Musunuri and Chapman 2005; Abdel-Rahman et al. 2008). Methods to reduce V_{swing} include statically (Kursun et al. 2004) and dynamically (Mulligan et al. 2005) reducing the gate drive voltage swing, compared to the otherwise rail-to-rail V_{swing}.

The second approach of improving light load efficiency focuses on reducing P_Q: the static or quiescent power consumption of the converter circuits. P_Q includes the quiescent power of circuits like voltage reference and biasing, comparators, and the leakage in the power MOSFETs, etc. In most applications where $I_{out} \geq 10\,mA$, P_Q would be minimal compared to other type of power losses. However, when I_{out} decreases to a point that it becomes comparable to I_Q, then P_Q becomes the limitation.

Xiao et al. (2004) presented a dual-mode all-digital PWM/PFM buck converter that reduced its I_Q to only $4\,\mu A$. The converter achieved 70% power efficiency at $I_{out} = 100\,\mu A$ in $0.25\,\mu m$ standard CMOS process. Notice that the regular output power range for the buck converter was 0-400 mA, 1.2 V. Ramadass and Chandrakasan (2008) proposed approximate zero-current switching, which avoids using power hungry high-gain amplifiers and complicated control circuits for precise zero-current detection. As a result, the quiescent current is greatly reduced, and it achieved above 80% efficiency from 1 to $100\,\mu A$ I_{out} in 65 nm CMOS process.

Another common method of reducing P_Q is "burst mode" operation (Wilcox and Flatness 2003; Salerno and Jordan 2006; Chen 2007). It is based on the observation that not all parts of the converter need to be on all the time, especially during light load. The power MOSFETs deliver energy to the output until the output voltage is within the pre-determined range. Since the load is light, the output capacitor alone will be able to hold the voltage for an extended period of time. As long as the output voltage is within the preset boundaries, non-essential parts of the converter are turned off, making the whole chip consuming less quiescent power.

Assuming that the wake time ("burst") and the sleep time (no "burst") of a DC-DC converter are T_{wake} and T_{sleep} respectively, and the converter's power consumption during sleep time is negligible, the equivalent quiescent current $I_{Q,eff}$ of the converter can be found as:

$$I_{Q,eff} = I_{Q,wake} \cdot \frac{T_{wake}}{T_{wake} + T_{sleep}} \qquad (4.4)$$

From (4.4), the longer T_{sleep} is compared to T_{wake}, the lower $I_{Q,eff}$ can be. As a result, P_Q can be reduced indirectly by controlling the sleep-wake duty cycle $\mathcal{D}$

$$\mathcal{D} = \frac{t_{wake}}{T} = \frac{t_{wake}}{t_{wake} + t_{sleep}} \qquad (4.5)$$

where T is the sleep-wake period, or the inverse of "burst period" ($T = 1/f_{burst}$). Compared to other methods of directly reducing the quiescent current I_Q (Xiao et al. 2004; Ramadass and Chandrakasan 2008), "burst mode" operation do not need to actually scale down the quiescent current for circuits, which can often lead to undesirable performance penalties or increase in die area.

However, the sleep-wake ratio $\mathcal{D}$ can not be set arbitrarily. Too small a sleep-wake ratio may lead to large inductor current ripple that would degrade the efficiency. The minimum $\mathcal{D}$ is also a function of load current and output capacitor size, as a larger output capacitor would be able to hold the voltage for a longer period of time.

As a result, many "burst mode" DC-DC converter do not actively control $\mathcal{D}$, but uses hysteretic mode control for voltage regulation and peak inductor current control to implement the bursts. If the output voltage drops below the lower threshold V_{th-}, the high-side MOSFET is turned on such that the inductor current rises. The high-side MOSFET is turned off and low-side MOSFET is turned on after the inductor current reaches a pre-set value $I_{L,peak}$. The inductor current then decreases until it reaches zero again. Output voltage V_{out} is checked at this point. If it is below the upper threshold V_{th+}, then another fixed-$I_{L,peak}$ switching period is issued. If V_{out} is above V_{th+}, then the converter no longer switches and goes into standby (Chen 2007).

A further study of "burst mode" operation would reveal that it resembles a charger. A non-bursting conventional DC-DC converter operates continuously, meaning that the apparent output power of the converter, P_{out}, is always equal to the required load power, P_{load} in the steady state. A "burst mode", or sleep-wake DC-DC converter, on the other hand, breaks the connection between P_{out} and P_{load}. The apparent output power of the converter P_{out} during "bursts" are set to be higher than P_{load} such that the output voltage is charged up after a series of "bursts". It is also because of this charging that no-burst idle time became possible.

Convenient as hysteretic and peak inductor current controls are in Chen (2007), they are not necessarily the most power efficient implementation of "burst mode" operation. Jang and Jovanovic (2010) proposed implementing "burst" from a different angle: the apparent output power $P_{out,app}$ during t_{wake} should match the output power at which the converter achieves the maximum efficiency. In this book, we refer to it as the principle of power matching.

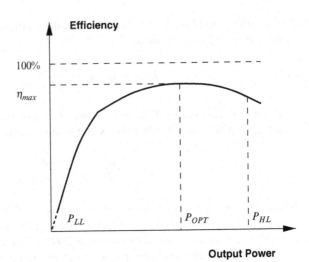

Fig. 4.3 A general illustration of the optimum output power point P_{opt} at which a DC-DC converter achieves maximum efficiency η_{max}

Notice that every DC-DC converter reaches its intrinsic maximum power efficiency (η_{max}) at an optimum output power point, P_{opt}, as seen in Fig. 4.3. In Jang and Jovanovic (2010), the wake-time apparent output power $P_{out,app}$ is chosen to be P_{opt}, as the sleep-time apparent output power is assumed to be zero. In this way, the maximum overall efficiency of η_{max} can still be achieved ideally.

Because the average power delivered to the load needs to be P_{load} regardless of how $P_{out,app}$ is chosen, a proper sleep-wake ratio $\mathscr{D}$ should also be selected. To put things in perspective, a conventional DC-DC converter without "burst mode" operation has $\mathscr{D} = 1$, i.e. always awake.

$$\mathscr{D} = \frac{P_{load}}{P_{out,app}} \equiv 1 \tag{4.6}$$

$$\eta = \frac{P_{out,app}}{P_{in}} = \frac{P_{load}}{P_{in}} \leq \eta_{max} \tag{4.7}$$

"Burst mode" operation, on the other hand, allows

$$\mathscr{D}_{opt} = \frac{P_{load}}{P_{opt}} < 1 \tag{4.8}$$

$$\eta_{opt} = \frac{P_{out,app}}{P_{in}} = \frac{P_{opt}}{P_{in}} = \eta_{max} \tag{4.9}$$

where $\mathscr{D}_{opt}$ is the corresponding optimum sleep-wake ratio as a result of the principle of power matching.

Another major advantage of the method in Jang and Jovanovic (2010) over previous solutions (Zhou et al. 2000; ho Choi et al. 2004; Zhou and Rincon-Mora 2006; Abdel-Rahman et al. 2008) is that it allows the reuse of existing heavy-load

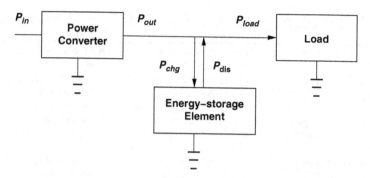

Fig. 4.4 Block diagram of a Sleep-Wake DC-DC converter with an energy storage element

efficient DC-DC converters for light load operations, avoiding the circuit re-design and optimization for each new load condition. This is because existing power converter IPs for portable electronics are usually optimized for normal mode, where $\eta_{existingIP} \approx \eta_{max}$, making the whole converter design directly re-usable for light-load with only some added sleep-wake control.

A key limitation of Jang and Jovanovic (2010), however, is the power loss associated with charging and discharging the energy storage elements. Some loads cannot handle discontinuous batches of power. As a result, Some proper energy storage elements are needed to smooth out the power transfer. Figure 4.4 shows the block diagram of a sleep-wake DC-DC converter with an energy storage element, which absorbs the excess power $P_{out} - P_{load}$ during wake time and discharge itself to provide for P_{load} during sleep time. The charge (P_{chg}) and discharge (P_{dis}) power of the energy storage element can be found as

$$P_{chg} = P_{out} - P_{load} \qquad (4.10)$$

$$P_{dis} = P_{load} \qquad (4.11)$$

Assume that the charge and discharge efficiencies are η_c and η_d respectively, based on the conservation of power

$$\eta_c \cdot \eta_d \cdot P_{chg} \times t_{wake} = P_{dis} \times t_{sleep} \qquad (4.12)$$

Thus, the wake-sleep duty ratio and the overall efficiency with charging and discharging loss included can be calculated as

$$\mathscr{D} = \frac{1}{(1 - \eta_{ES}) + \eta_{ES}(P_{out}/P_{load})} \qquad (4.13)$$

$$\eta = \frac{P_{load} \times (t_{wake} + t_{sleep})}{P_{in} \times t_{wake}} = \frac{P_{load}}{P_{out}} \cdot \frac{P_{out}}{P_{in}} \cdot \frac{1}{\mathscr{D}}$$

$$= \eta_{wake} \frac{\eta_{ES}}{1 - \mathscr{D}(1 - \eta_{ES})} \qquad (4.14)$$

Table 4.1 Advantages and disadvantages of existing light-load efficiency boosting techniques

Technique	Advantages	Disadvantages
Variable f_s (Arbetter et al. 1995)	Efficiency improvement independent of load	DCM only, output spectrum needs to be randomized
Hybrid mode (Zhou et al. 2000)	Flexibility in selecting sync. and async. rectifier for optimum efficiency	Trade-off between conduction loss (Schottky) and switching loss
Traditional mode-hopping (Zhou and Rincon-Mora 2006)	Flexibility in selecting CCM/DCM for optimum efficiency	Performance degrades with large output voltage ripple
Improved mode-hopping (Qahouq et al. 2007)	Constant ripple and fast transient with good efficiency	Trade-off between efficiency and performance
Non-linear inductor (Sun et al. 2009)	Different inductance value for PFM/PWM	Usage of non-standard component: saturable inductor
Gate drive scaling (Mulligan et al. 2005; Kursun et al. 2004)	Reduces switching loss	Limited effectiveness due to other power loss factors
I_Q reduction (Xiao et al. 2004)	Improved light load efficiency	Difficulties in designing low I_Q circuitry
"Burst mode" operation (Wilcox and Flatness 2003; Salerno and Jordan 2006)	High efficiency over broad current range	Efficiency limited by always-on circuitry
Improved "burst mode" (Jang and Jovanovic 2010)	Maximum efficiency from medium to light load range	Efficiency limited by energy storage elements. P_{opt} not known *a priori*

where $\eta_{ES} = \eta_c \cdot \eta_d$ is the combined efficiency of the energy storage element. In the ideal case of $\eta_{ES} = 1$, (4.13) and (4.14) become (4.8) and (4.9).

Another drawback of Jang and Jovanovic (2010)'s method is the fact that P_{opt} is not known *a priori*: it depends on the design parameters and the physical implementation of the DC-DC converter. In other words, η_{max} and P_{opt} can only be determined or measured after the converter is designed and implemented, and their values may depend heavily on the components and topologies chosen. As a result, the sleep-wake ratio $\mathscr{D}$ is often picked according to experience (Jang and Jovanovic 2010) instead of the design equation of (4.13). Finally, a summary of existing light-load efficiency improvement methods and their advantages and disadvantages are listed in Table 4.1.

4.3 The Long-Sleep Model

The improved sleep-wake method with the principle of power matching (Jang and Jovanovic 2010) is particularly suitable for emerging multiple power mode electronics with emphasis on sleep-mode efficiency. The low cost nature of certain products even prohibits fully-customized design for each additional power saving mode, but the design and power consumption for normal mode is well defined.

If an existing IP of a converter is optimized for the active mode, then $P_{load,active}$ should be very close to P_{opt} already. In order to provide a much lighter $P_{load,sleep}$, the wake-time output apparent power $P_{out,wake}$ should be set to $P_{load,active}$, and the resulting sleep-wake duty cycle $\mathscr{D}$ will be determined by the ratio of $P_{load,active}$ and $P_{load,sleep}$. In this way, $\mathscr{D}$ should be very close to $\mathscr{D}_{opt}$, and the intrinsic maximum efficiency η_{max} could still be achieved.[3]

Due to the huge difference between $P_{load,active}$ and $P_{load,sleep}$ in many applications For example, the supply current of a coin-cell-powered wireless sensor can vary from tens of mA during radio transmission (TX) and reception (RX) to a few μA in sleep (Cook et al. 2006; TIC 2009), the sleep time could be significantly longer than the wake time, leading to a Long-Sleep Model (LSM) with certain unique characteristics (Hu and Ismail 2011), which will be described in this section.

4.3.1 Definition

The Long-Sleep Model (LSM) describes a sleep-wake DC-DC converter with significantly longer sleep time than wake time. In this book, significantly long refers to a ratio of or more than three orders of magnitudes.[4]

$$\frac{t_{sleep}}{t_{wake}} \geq 10^3 \tag{4.15}$$

If we further define $T = t_{sleep} + t_{wake}$ as the sleep-wake period, then the LSM refers to scenarios with $\mathscr{D} < 10^{-3}$ and $t_{sleep} \approx T, t_{wake} = \mathscr{D}T$.

4.3.2 Implication: Large-I_0 Approximation

The first implication of LSM is the Large-I_0 approximation . Let I_0 be the peak inductor current during t_{wake}. Since $t_{sleep} \gg t_{wake}$, I_0 is much bigger than I_{load}. More specifically, Large-I_0 approximation states that

$$I_0 \gg I_{load} \text{ s.t. } I_{load} = 0 \text{ during } t_{wake} \tag{4.16}$$

---

[3]To be more precise, only η_{active} is achieved. But η_{active} should be very close to, if not identical to η_{max}, assuming perfect design and optimization for the active mode application.

[4]The choice of 10^3 within the definition (4.15) is relatively arbitrary. A less pronounced difference in t_{wake} and t_{sleep} could make some of the conclusions in this chapter less accurate. Ultimately, it is at the discretion of the designer to decide whether an LSM-approximated analysis and results are accurate enough for their applications.

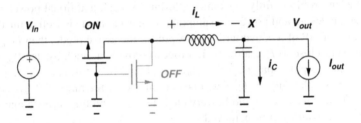

Fig. 4.5 Large-I_0 approximation in a typical buck converter

The importance of Large-I_0 approximation manifests itself when applying KCL at the node X in a typical buck converter as seen in Fig. 4.5.

$$i_L = i_C + I_{load} \approx i_C \qquad (4.17)$$

If we integrated both sides of (4.17) over one inductor current period, we have

$$\Delta Q_L = \Delta Q_C \qquad (4.18)$$

which reveals the underlying principle of charger conservation for large-I_0 approximation: any electric charge delivered through the inductor L, which ultimately came from the battery, is stored in the capacitor C, as I_{load} is too small compared to I_L during the wake-up moment.

4.3.3 Characteristics

4.3.3.1 Inductor Current

The first implication of a much longer sleep is the "pulsification" of inductor current (Fig. 4.6). Throughout the sleep time, there is no power delivered to the output. Thus, the inductor current I_L is zero. The time when the inductor current is not zero (such as during t_{wake}) is so short that they look like pulses.

Within the "pulsified" t_{wake}, there are still three possible I_L scenarios denoted as Type I, II, and III, as seen in Fig. 4.6. In the Type I scenario, the buck converter operates in a manner similar to continuous conduction mode (CCM), except for the initial ramp-up and final ramp-down phases. The converter can also operate on the boundary between continuous conduction mode (CCM) and discontinuous conduction mode (DCM) (Type II) or in DCM completely (Type III).

The principle of power matching would determine which form the inductor current will take. The principle of power matching states that $P_{out,app} = P_{opt}$ during

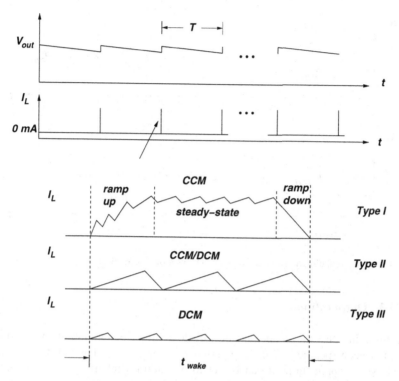

Fig. 4.6 Long-Sleep model waveforms and inductor current profiles

t_{wake}. Since a generic DC-DC converter is more likely to achieve its best efficiency in CCM or CCM/DCM boundary than in DCM, the inductor current I_L is more likely to look like Type II and III than Type I.

4.3.3.2 Load Regulation

In applications where the input and output voltages of the DC-DC converter are constant, the sleep-wake duty cycle $\mathscr{D}$ is an indicator of the output power level and the counterpart of the average inductor $I_{L,avg}$ of a conventional CCM DC-DC converter. Therefore, the two topologies respond to a load transient differently.

When there is a load increase from I_{load1} to I_{load2} (where V_{out} is constant), $\mathscr{D}_1$ needs to be adjusted to a new value $\mathscr{D}_2$, meaning the converter needs to wake up more often (as depicted in Fig. 4.7a), but the duty cycle D during t_{wake} may or may not be changed. In a conventional DC-DC converter, however, the average inductor current $I_{L,avg1}$ will increase to $I_{L,avg2}$, which is achieved by a temporary transient (simplified as "$D = 1$" in Fig. 4.7b). The duty cycle $D = D_0$ remains the same before and after the new steady-state is established.

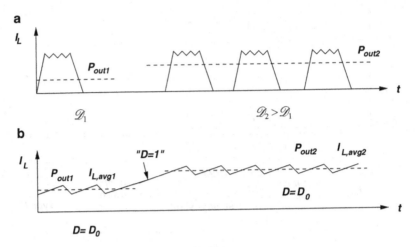

Fig. 4.7 Load regulation of a conventional DC-DC vs a sleep-wake DC-DC

4.3.3.3 Output Ripple

Till this point, only the sleep-wake duty cycle $\mathcal{D}$ and apparent output power during t_{wake} has been specified. The sleep period $T = t_{sleep} + t_{wake}$ has not been chosen, and in some applications it can be set to be within a relatively wide frequency range (Jang and Jovanovic 2010). However, it is important to notice that a group of performance parameters, such as the output ripples, minimum recovery time, off-chip component size, and achievable efficiency in practice are all related to T.

The output ripple of a DC-DC converter can be expressed as

$$\Delta V_{out} = \Delta V_C + \Delta V_R \tag{4.19}$$

where ΔV_C and ΔV_R refers to the ripple components caused by the capacitor and its ESR respectively. In a sleep-wake DC-DC converter, the energy storage element C would discharge itself to provide the output current during t_{sleep}. Therefore,

$$I_{load} = C \cdot \frac{\Delta V_C}{T} \tag{4.20}$$

$$\therefore \Delta V_C = \frac{I_{load} \cdot T}{C} \tag{4.21}$$

The shorter the sleep-wake period, the smaller the ripple. From an efficiency perspective, it is desirable to have longer T to reduce the switching and its associated power loss, but it also increases ΔV_C. A closer study would show that a longer T would also increase ΔV_R, as $\Delta V_R \approx I_0 R_{ESR}$ (according to the Large-I_0 approximation: $i_L \approx i_C$ during t_{wake}). As T increases, $\Delta Q_C = C \Delta V_C$ increases, whereas

$$I_0 = \frac{2}{t_{wake}} \int_0^{t_{wake}} i_L(t)\, dt = \frac{2\Delta Q_L}{t_{wake}} = \frac{2\Delta Q_C}{t_{wake}} \tag{4.22}$$

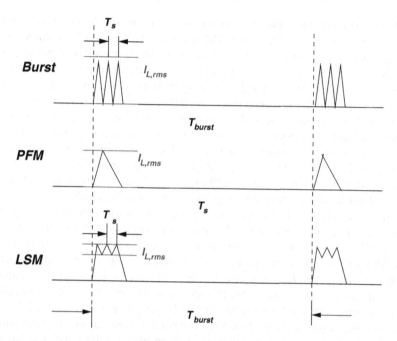

Fig. 4.8 Comparison of the inductor current waveform of burst mode operation, pulse-frequency modulation and the long-sleep model

Thus, ΔV_R would also increase as a result. In summary, longer T reduces P_{sw} but increases the inductor ripple, which could potentially increase P_{cond}, as analyzed in Sect. 4.2. The proper sleep-wake period T will need to be chosen based on the efficiency, ripple requirements, and external component sizes.

4.3.4 Novelty

The LSM model proposed here is similar in many ways to the conventional pulse-frequency modulation (PFM) (Sahu and Rincon-Mora 2007) and "burst mode" operation (Wilcox and Flatness 2003; Salerno and Jordan 2006; Chen 2007; Jang and Jovanovic 2010). Their similarities and differences are best explained by Fig. 4.8, which shows the inductor current waveforms of the three methods.

The three methods are similar in that the inductor currents stay at zero for an extended period of time each switching or bursting cycle. It is because of these idling time that the switching related power loss P_{sw} can be reduced, which improves the light-load efficiency substantially.

The first difference is the cause for zero inductor current. In PFM, zero I_L is a result of synchronous rectification and reverse inductor current prevention in the discontinuous-conduction mode (DCM) operation. In burst-mode operations, including both the conventional bursts and the LSM, the zero-I_L period is the result of hysteretic voltage control, meaning that the output voltage is still considered high enough that no additional current, or ultimately power, is needed from the source.

The second difference is what happens during $I_L = 0$. In a regular PFM operation, there is no specific power saving feature other than the negative I_L detection for synchronous rectification. Circuit blocks are left idle and consume quiescent current as usual. In burst-mode and LSM, however, an effort was made to partially or completely turn off unnecessary circuitry to save on quiescent power P_Q. This is because the burst mode operation and the LSM are specifically designed to reduce P_Q as part of the efficiency boosting strategy. This partial shut-down strategy is also more feasible in burst modes than in PFM, as t_{wake} and t_{sleep} are more predictable and well managed (for instance through hysteretic voltage control).

The third difference is the operation during t_{wake}, or when the inductor current is not zero. In PFM operation, synchronous rectification took place. The high side MOSFET is turned on first, allowing the inductor current to ramp up. Then the high-side MOSFET is turned off with the low-side MOSFET on, depleting the inductor current to zero. Once $I_L = 0$ is detected, both MOSFETs are switched off, and a high impedance node is seen by the inductor. In burst-mode operation, consecutive inductor current "bursts" are issued. The I_L bursts will not stop until the output voltage is high enough and the hysteretic comparator signals an end to the bursting period. In the LSM, a more sophisticated I_L pattern other than back-to-back "bursts" is present. I_L can take the form of Type I, II, and III profile, depending on the intrinsic maximum efficiency point (P_{opt}) of the power converter.

It is this difference in operation during t_{wake} that gives the LSM a potential advantage over the other two methods. First of all, the RMS current of the inductor $I_{L,rms}$ can be significantly smaller, leading to power savings in P_{cond}. Notice that the peak-to-peak inductor current is always approximately twice the average inductor current during t_{wake}. The RMS inductor current for LSM, on the other hand, is independent of $I_{L,avg}$ during t_{wake}, and it can be significantly smaller, as shown in Fig. 4.8. Secondly, the whole I_L profile during t_{wake} is matched to that of the converter at its intrinsic optimum efficiency point. Thus, $\eta_{wake} \approx \eta_{opt}$.

In summary, burst mode operation as depicted in the top trace of Fig. 4.8 represents a physical implementation-oriented approach. The LSM method, on the other hand, is a purely efficiency-oriented operation that ideally achieves the maximum efficiency across a wide range of output power. The physical design and implementation of an LSM DC-DC converter has its unique set of challenges, and a design example is given in the next section.

4.4 LSM Buck Converter Design Example

In this section, a design example of a dual power mode DC-DC buck converter using the LSM is presented. The buck converter is designed for a battery-powered micro-system similar to that of Fig. 1.4. In order to be cost effective, the buck converter

Table 4.2 Design goals for
the buck converter

	Active mode	Sleep mode
V_{in}	3 V	3 V
V_{out}	1.8 V	1 V
I_{out}	27 mA	5 μA
P_{out}	50 mW	50 μW

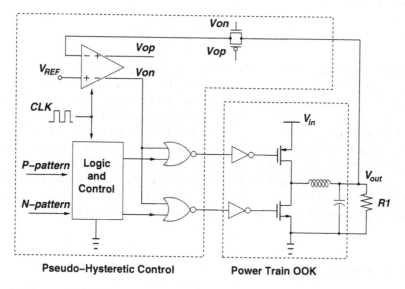

Fig. 4.9 System block diagram of the DC-DC converter

will be fully integrated with the rest of the system in a mainstream digital CMOS technology, except for an external capacitor (C) and inductor (L). To achieve multi-year operation for the system without battery replacement, the buck converter will have to be highly efficient in all power modes. For simplicity, this example assumes two power modes only. Table 4.2 lists the design goals for the buck converter.

4.4.1 System Design Considerations

Figure 4.9 shows the system block diagram of the DC-DC converter. The power train, which includes the power MOSFETs, gate driving buffers, external inductor (L) and capacitor (C) will be optimized for active mode $P_{out,active}$ such that $P_{opt} = P_{out,active}$ and $\eta_{active} = \eta_{max}$. The control loop, which includes the clocked comparator, transmission gate, and synchronization logic will be designed to obtain the appropriate sleep-wake duty ratio $\mathscr{D}$ of $\frac{P_{load,sleep}}{P_{load,active}}$ such that $\eta_{sleep} = \eta_{max}$.

4.4.2 Power Train Design

The power MOSFETs of the converters have to be sized appropriately to accommodate the maximum power rating and optimize the efficiency. The on-resistance of the power MOSFETs $R_{DS(ON)}$ in linear region can be reduced by increasing their aspect ratios W/L.

$$R_{DS(ON)} = \frac{1}{\mu_{p(n)} C_{ox} \frac{W}{L} (V_{GS} - V_{TH})} \tag{4.23}$$

On the other hand, large size MOSFETs have larger gate-source capacitance C_{gs}, which requires more gate driving power. Since $C_{gs} \propto W \cdot L$, the minimum channel length in the technology $L = 0.6\,\mu m$ was chosen to minimize C_{gs} while achieving the same low $R_{DS(ON)}$ values. In this design, a width of 9,000 μm was chosen for the NMOS to achieve a $R_{DS(ON)}$ of 440 $m\Omega$ with C_{gs} around 7 pF and C_{gd} of 6 pF. (Notice that the device is mostly in linear region.) A width of 60,000 μm was chosen for the PMOS to achieve an $R_{DS(ON)}$ of 180 $m\Omega$ with C_{gs} and C_{gd} around 55 pF and 54 pF respectively. Tapered CMOS inverters chains are added as gate drivers for both the PMOS and NMOS power transistors. A tapering ratio of ten is chosen as a compromise of speed and gate drive power (Cherkauer and Friedman 1995).

4.4.3 Control Loop Design

4.4.3.1 Hysteretic Control

There are a number of ways of designing the DC-DC control loop such that the output voltage is stable against load variation. Voltage mode control (VMC) is easy to implement but difficult to stablize or achieve a fast loop response (Liou et al. 2008; Sahu and Rincon-Mora 2007). Current mode control (CMC) provides simpler loop compensation and fast load regulation, but it requires accurate load current sensing and current feedback loop, which increases the converter complexity and power consumption (Lee and Mok 2004; Huang et al. 2007).

Hysteretic control (HC) is simple and inherently stable, i.e., it does not require loop compensation. It also responds to load variation quickly, and it is efficient for light-load operations (Su et al. 2008; Huang et al. 2009). The draw back of hysteretic control is its relative large output ripples and variable switching frequency, which could complicate Radio Frequency Interference (RFI) and Electromagnetic Interference (EMI) design.

The characteristics mentioned above make hysteretic control uniquely suitable for sleep-wake operation, though certain modifications are necessary. Output ripple is not as critical in sleep modes, as key analog and mixed-signal blocks are off, and the circuits that remain on enjoy much more relaxed performance requirements in supply variation, RFI, and EMI. Variable switching frequency, however, remains an issue, as timing accuracy in sleep mode is greatly reduced compared to active mode.

Hence, a clocked hysteretic control (CHC) is proposed. Instead of using a hysteretic comparator, whose output directly controls the on and off of the power MOSFETs, a clocked comparator is used to periodically sample the output voltage, V_{out}. If V_{out} falls below a certain level V_{ref}, the comparator signals a "wake" for the DC-DC converter. The converter power train will then operate in an open-loop fashion for a brief t_{wake}, during which the power MOSFETs are switched on and off according to a pre-determined pattern and a fixed P_{out} is delivered to the output, before going back to sleep. If V_{out} is above V_{ref}, the comparator sends a "sleep" signal. The DC-DC converter will then remain in sleep until the next sampling period.

As such, an equivalent hysteretic voltage window ΔV_{out} is created, the magnitude of which is related to the sampling clock period t_{wake} and P_{out}. According to the large-I_0 approximation of the LSM, $I_{out} \approx I_C$. Therefore

$$\Delta V_{out} = \frac{I_C \cdot t_{wake}}{C} \approx \frac{P_{out} t_{wake}}{V_{out} C} \tag{4.24}$$

Clocked comparison also allows for better synchronization with other system activities. In addition, using fixed pattern instead of the hysteretic comparator's output to switch on and off the power MOSFETs avoids variable switching frequency and RFI/EMI problems.

4.4.3.2 Clocked Comparator

The clocked comparator is the key block that implements the proposed CHC control. To reduce the power loss overhead introduced by the circuit, a zero-DC current comparator (Xiao et al. (2004)) is adopted, as seen in Fig. 4.10. Notice that the comparator in Fig. 4.10. is a level sensitive latch. To ensure that the hysteretic window is created properly, and that the DC-DC converter operates in a pseudo open-loop fashion during t_{wake}, a CMOS transmission gate is added in the feedback loop, as seen in the block diagram in Fig. 4.9, effectively making the comparator an edge sensitive latch.

4.4.3.3 Gate Drive Patterns and Logic

The pre-determined gate drive patterns during t_{wake} are designed to match the operation at P_{opt}. As explained in Sect. 4.3.3.1, the Type-I scenario, which is CCM operation during t_{wake}, is most likely. In this example, a duty cycle operation profile was selected as one possible design whose P_{opt} coincides with $P_{load,active}$. Its synchronous rectification and optimum deadtime control are all embedded into the gate drive pattern. Once designed, the patterns can be stored in system ROM, RAM, or registers for easy access.

Fig. 4.10 Schematic of the
clocked comparator

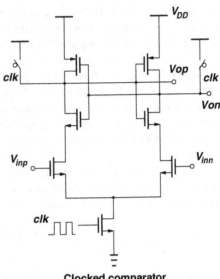

Clocked comparator

The remaining logic circuits in Fig. 4.9 control the sleep-wake transition. The output of the clocked comparator enables or disables the gate drive pattern through CMOS NOR gates, which effectively turns the power train on and off for wake and sleep.

4.4.4 Simulation Results

The LSM buck converter design example was implemented in MOSIS 0.5 μm digital CMOS technology. The test benches in normal and sleep modes are shown in Fig. 4.11. Open-loop simulation was performed to mimic normal mode operation and guide the power train design to have its P_{opt} matched with $P_{load,active}$. Close-loop simulation was performed in sleep mode to check the load regulation and efficiency. External stimuli include the voltage reference, wake-up clock, pre-determined NMOS and PMOS gate drive signal patterns.

The open-loop simulation of the optimized power train was shown in Fig. 4.12 and 4.13. The loop dynamic is shown in Fig. 4.12, where V_{out} starts from an initial voltage and settles to a pre-determined final value of 1 V (1.03 V in simulation) within 50 μs. Since it is an open-loop simulation, V_{out} is not regulated against output load, but it will change with different duty cycles reflected in different gate driving patterns. The detailed inductor current I_L waveform is shown in Fig. 4.13, where it can be seen that the optimized power train operates on the CCM-DCM boundary with peak I_L of 100 mA and switching frequency of 1.37 MHz. The efficiency η_{max}, which is the ratio of the power delivered to R_{out} in Fig. 4.11 over the power drained from the battery V_{in}, is measured to be 89%.

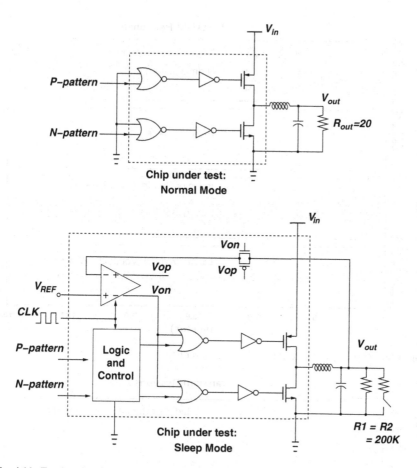

Fig. 4.11 Test benches for the DC-DC converter in normal mode and sleep mode

Figure 4.14 shows the close-loop sleep-wake operation of the buck converter. The top trace in Fig. 4.14 is the sleep timer available in the system. Depending on the actual load, the DC-DC converter wakes up every several clock cycles. The inductor current was "pulsified" as expected from analysis in Sect. 4.3.3.1, indicating the appropriateness for various LSM approximations. (Equation 4.15). During t_{wake}, the inductor current matches that of Fig. 4.13, as illustrated in Fig. 4.15, which means the apparent output power during t_{wake} has been matched to P_{opt} for maximum efficiency.

The transient load regulation of the buck converter was shown in Fig. 4.16. An abrupt, 100% load increase from $5\,\mu A$ to $10\,\mu A$ was simulated by a switching on a $R_2 = 200K$ in parallel with R_1 as seen in Fig. 4.11. The converter output voltage experienced a temporary dip, and the power train waked up more frequently, i.e. on every wake-up clock cycle. Consequently, V_{out} recovers within five sampling cycles (T_s), or 50 ms. The significance of achieving good load regulation in sleep mode is

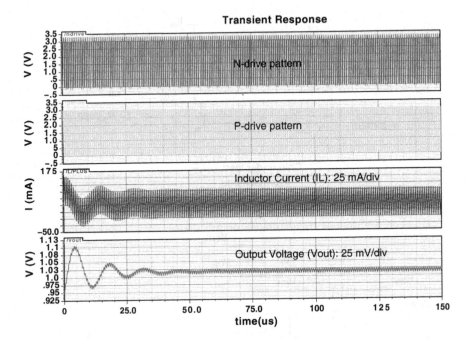

Fig. 4.12 Open-loop dynamic of the optimized power train

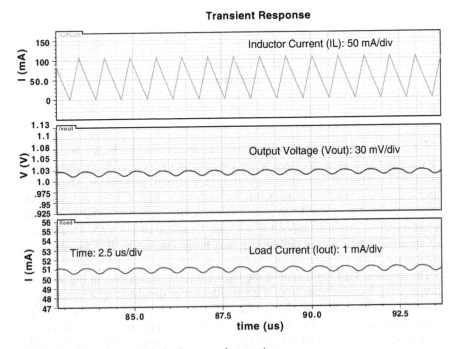

Fig. 4.13 Inductor current I_L during heavy mode operation

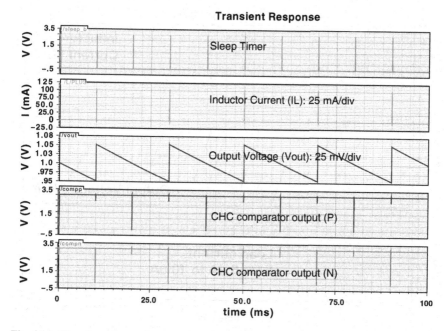

Fig. 4.14 Close-loop sleep-wake operation of the buck converter

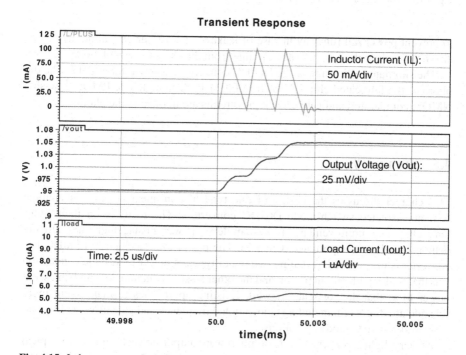

Fig. 4.15 Inductor current I_L during t_{wake}

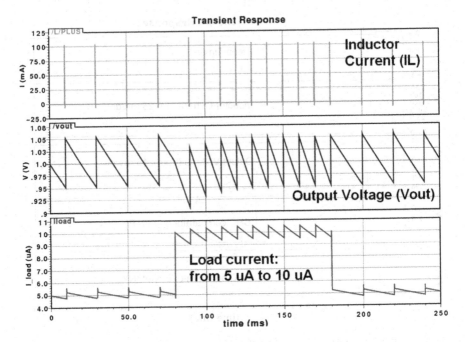

Fig. 4.16 Load regulation in sleep mode

to prevent power rail runaway in case of leakage and off-current variations, which also helps reduce the system wake-up time from the sleep to normal mode.

The simulated efficiency of the DC-DC converter is shown in Fig. 4.17. The die photo of the fabricated chip is shown in Fig. 4.18. The high-side PMOS, low-side NMOS, comparator, and gate drive logic circuits are marked in the photo.

4.5 Summary

This chapter discusses the design of one type of non-linear power management ICs: non-isolated switch-mode DC-DC converters. The focus is on their light-load efficiency improvement, as system level power management has already demonstrated their importance (Chap. 2). The power loss mechanism of a DC-DC converter is first studied, revealing two major types of power losses: the conduction loss that scales with output power, and the other losses (switching loss and quiescent power loss) that do not scale as well. Existing approaches to make them more scalable are discussed, including conventional frequency and gate drive scaling and sleep-wake "burst" mode operations.

At very light loads (milliwatt or microwatt output power), which are typical for the sleep mode power consumptions of multi-year battery-powered devices,

Fig. 4.17 Simulated
efficiency of the proposed
converter

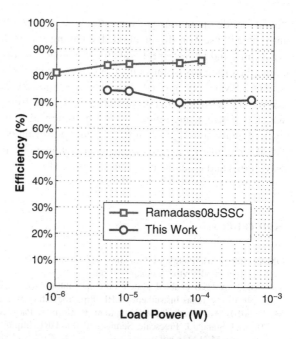

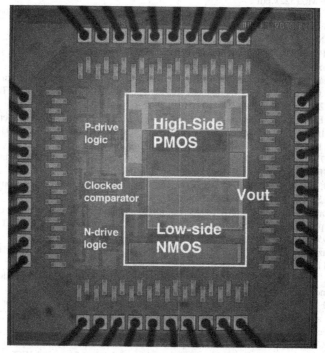

Fig. 4.18 The die photo of the LSM buck converter

"burst" mode operations are very attractive. A novel long-sleep model (LSM) was introduced based on this method. By having a very large sleep-wake ratio and a wake time inductor current waveform matched to that of the optimum operating point, the efficiency of the converter can be maintained across a wide load range.

The LSM can be applied to the design of multiple power mode efficient DC-DC converters. It enables maximum silicon IP reuse from active mode to sleep mode, which will effectively reduce the design time and shorten the product development cycle. Finally, a design example of a dual power mode, microwatt sleep mode DC-DC Buck converter is provided featuring clocked hysteretic control. The simulated waveforms and efficiencies are also presented.

References

OMA (2008) OMAP3430 Multimedia Applications Processor. Texas Instruments, URL http://focus.ti.com/pdfs/wtbu/ti_omap3430.pdf

TIC (2009) A USB enabled system-on-chip solution for 2.4-GHz IEEE 802.15.4 and Zigbee Applications. Texas Instruments, URL http://focus.ti.com/lit/ds/symlink/cc2531.pdf

MC1 (2010) Advanced Zigbee-compliant Platform-in-Package (PiP) for the 2.4 GHz IEEE 802.15.4 Standard. Freescale Semiconductor, URL http://www.freescale.com/files/rf_if/doc/data_sheet/MC1322x.pdf

STM (2010) High-performance, IEEE 802.15.4 wireless system-on-chip. STMicroelectronics, URL http://www.st.com/stonline/products/literature/ds/16252/stm32w108cb.pdf

NRE (2010) National renewable energy laboratory. URL www.nrel.gov

TPS (2010) Power Management IC for Li-Ion Powered Systems. Texas Instruments, URL http://focus.ti.com/lit/ds/symlink/tps65023.pdf

Abdel-Rahman O, Abu-Qahouq J, Huang L, Batarseh I (2008) Analysis and design of voltage regulator with adaptive FET modulation scheme and improved efficiency. IEEE Trans Power Electron 23(2):896–906, DOI 10.1109/TPEL.2007.915184

Arbetter B, Erickson R, Maksimovic D (1995) DC-DC converter design for battery-operated systems. In: PESC1995: 26th IEEE Power Electron. Specialists Conf., vol. 1, pp 103–109 vol.1, DOI 10.1109/PESC.1995.474799

Chen J (2007) Determine buck converter efficiency in pfm mode. Tech. rep., National Semiconductor, URL http://powerelectronics.com/mag/709PET22.pdf

Cherkauer B, Friedman E (1995) A unified design methodology for cmos tapered buffers. IEEE Trans VLSI Syst 3(1):99–111, DOI 10.1109/92.365457

ho Choi J, young Huh D, seok Kim Y (2004) The improved burst mode in the stand-by operation of power supply. In: Proc. IEEE Appl. Power Electron. Conf., pp 426–432, DOI 10.1109/APEC.2004.1295844

Cook B, Lanzisera S, Pister K (2006) SoC issues for RF smart dust. Proc IEEE 94(6):1177–1196, DOI 10.1109/JPROC.2006.873620

Erickson R, Maksimović D (2001) Fundamentals of power electronics, 2nd edn. Springer Netherlands

Hu J, Ismail M (2011) A design method for light-load efficient dc-dc converter for low power wireless applications. In: Government Microcircuit Applications and Critical Technology Conference (GOMACTech), accepted for publication

Huang HH, Chen CL, Chen KH (2009) Adaptive window control (AWC) technique for hysteresis DC-DC buck converters with improved light and heavy load performance. IEEE Trans Power Electron 24(6):1607–1617, DOI 10.1109/TPEL.2009.2014687

Huang HW, Chen KH, Kuo SY (2007) Dithering skip modulation, width and dead time controllers in highly efficient DC-DC converters for system-on-chip applications. IEEE J Solid-State Circuits 42(11):2451–2465, DOI 10.1109/JSSC.2007.907175

Jang Y, Jovanovic M (2010) Light-load efficiency optimization method. IEEE Trans Power Electron 25(1):67–74, DOI 10.1109/TPEL.2009.2024419

Kursun V, Narendra S, De V, Friedman E (2004) Low-voltage-swing monolithic DC-DC conversion. IEEE Trans Circuits Syst II, Express Briefs 51(5):241–248, DOI 10.1109/TCSII.2004.827557

Lee CF, Mok P (2004) A monolithic current-mode cmos dc-dc converter with on-chip current-sensing technique. IEEE J Solid-State Circuits 39(1):3–14, DOI 10.1109/JSSC.2003.820870

Liou WR, Yeh ML, Kuo YL (2008) A high efficiency dual-mode buck converter ic for portable applications. IEEE Trans Power Electron 23(2):667–677, DOI 10.1109/TPEL.2007.915047

Ma D, Ki WH, Tsui CY (2004) An integrated one-cycle control buck converter with adaptive output and dual loops for output error correction. IEEE J Solid-State Circuits 39(1):140–149, DOI 10.1109/JSSC.2003.820844

Mohan N, Undeland TM, Robbins WP (2003) Power Electronics: Converters, Applications and Design, 3rd edn. John Wiley & Sons, Inc.

MOSIS (2011) On semiconductor c5 process. URL http://www.mosis.com/on_semi/c5/

Mulligan M, Broach B, Lee T (2005) A constant-frequency method for improving light-load efficiency in synchronous buck converters. IEEE Power Electron Lett 3(1):24–29, DOI 10.1109/LPEL.2005.845177

Musunuri S, Chapman P (2005) Improvement of light-load efficiency using width-switching scheme for cmos transistors. IEEE Power Electron Lett 3(3):105–110, DOI 10.1109/LPEL.2005.859769

Qahouq J, Abdel-Rahman O, Huang L, Batarseh I (2007) On load adaptive control of voltage regulators for power managed loads: Control schemes to improve converter efficiency and performance. IEEE Trans Power Electron 22(5):1806–1819, DOI 10.1109/TPEL.2007.904232

Ramadass Y, Chandrakasan A (2008) Minimum energy tracking loop with embedded dc-dc converter enabling ultra-low-voltage operation down to 250 mv in 65 nm cmos. IEEE J Solid-State Circuits 43(1):256–265, DOI 10.1109/JSSC.2007.914720

Sahu B, Rincon-Mora G (2007) An accurate, low-voltage, cmos switching power supply with adaptive on-time pulse-frequency modulation (pfm) control. IEEE Trans Circuits Syst I, Reg Papers 54(2):312–321, DOI 10.1109/TCSI.2006.887472

Salerno DC, Jordan MG (2006) Methods and circuits for programmable automatic burst mode control using average output current

Su F, Ki WH, Tsui CY (2008) Ultra fast fixed-frequency hysteretic buck converter with maximum charging current control and adaptive delay compensation for dvs applications. IEEE J Solid-State Circuits 43(4):815–822, DOI 10.1109/JSSC.2008.917533

Sun J, Xu M, Ren Y, Lee F (2009) Light-load efficiency improvement for buck voltage regulators. IEEE Trans Power Electron 24(3):742–751, DOI 10.1109/TPEL.2008.2009986

Wilcox ME, Flatness RG (2003) Control circuit and method for maintaining high efficiency over broad current ranges in a switching regulator circuit

Xiao J, Peterchev A, Zhang J, Sanders S (2004) A 4-ua quiescent-current dual-mode digitally controlled buck converter IC for cellular phone applications. IEEE J Solid-State Circuits 39(12):2342–2348, DOI 10.1109/JSSC.2004.836353

Zhou S, Rincon-Mora G (2006) A high efficiency, soft switching DC-DC converter with adaptive current-ripple control for portable applications. IEEE Trans Circuits Syst II, Express Briefs 53(4):319–323, DOI 10.1109/TCSII.2005.859572

Zhou X, Donati M, Amoroso L, Lee F (2000) Improved light-load efficiency for synchronous rectifier voltage regulator module. IEEE Trans Power Electron 15(5):826–834, DOI 10.1109/63.867671

Chapter 5
Conclusion

As the society at large faces the challenge to better utilize renewable energy sources and contain the environmental crises caused by carbon emission, improving energy efficiency in all aspects of lives is probably the most effective method to alleviate the problem. In consumer electronics, short battery life is inconvenient for customers. Better energy efficiency is becoming a must for future successful products.

In face of these challenges, this book introduces green electronics, a new class of energy-efficient portable and battery-powered electronic systems, as a potential solution. The concept of green electronics is introduced, together with discussions on its potential applications and possible architecture. Power management, an informed and intelligent effort to reduce power consumption, is then discussed through a holistic solution involving multiple levels of abstractions. Transistor-level power management IC designs, including capacitor-free low drop-out regulators and light-load efficient DC-DC converters, are discussed in details, following the logical sequence of explaining the performance requirements to reveal the technical challenges, analyzing the existing methods to understand their respective advantages and drawbacks, and proposing new solutions to better address the trade-offs.

Through these discussions, it was found that better energy efficiency can indeed be achieved in portable and battery-powered applications through various efforts. At the system level, a shift of focus from heavy load normal operation efficiency to light load standby efficiency boost could yield significant battery life improvement, as more portable battery-powered devices spend longer time in standby than in active usage. At the circuit level, sleep-mode efficient linear and switching power converter IPs can also be designed with techniques such as input current-differencing, excessive gain reduction, and the long-sleep model.

The book is characterized by a common theme: high efficiency and full on-chip integration, both of which stem from the unique requirements for portable and battery-powered applications, as well as the practical concern of manufacture-ability and cost. It is the first book on power management IC design with this focus.

J. Hu and M. Ismail, *CMOS High Efficiency On-chip Power Management*, Analog Circuits and Signal Processing, DOI 10.1007/978-1-4419-9526-1_5,
© Springer Science+Business Media, LLC 2011

Finally, the book does not seek, nor will it be *the* solution to the power management challenges. Rather, it is the authors' humble hope that this book can be a promising starting point for more scholarly research and industrial development on the same topic. Future research directions may include interface design for energy-harvesting devices, which have demonstrated capabilities to scavenge ambient thermal, kinetic, and electromagnetic energy at sub-mW levels to partially or fully power certain low-power electronics. Another promising angle would be investigating the power conditioning needs at the output of renewable energy generators, such as the output of mini-wind turbines and low-voltage solar cells, and design appropriate green electronics silicon IPs or SoCs to address those needs.

Index

J. Hu and M. Ismail, *CMOS High Efficiency On-chip Power Management*, Analog
Circuits and Signal Processing, DOI 10.1007/978-1-4419-9526-1,
© Springer Science+Business Media, LLC 2011